Lecture Notes in Networks and Systems 1257

The series "Lecture Notes in Networks and Systems" publishes the latest developments in Networks and Systems—quickly, informally and with high quality. Original research reported in proceedings and post-proceedings represents the core of LNNS.

Volumes published in LNNS embrace all aspects and subfields of, as well as new challenges in, Networks and Systems.

The series contains proceedings and edited volumes in systems and networks, spanning the areas of Cyber-Physical Systems, Autonomous Systems, Sensor Networks, Control Systems, Energy Systems, Automotive Systems, Biological Systems, Vehicular Networking and Connected Vehicles, Aerospace Systems, Automation, Manufacturing, Smart Grids, Nonlinear Systems, Power Systems, Robotics, Social Systems, Economic Systems and other. Of particular value to both the contributors and the readership are the short publication timeframe and the world-wide distribution and exposure which enable both a wide and rapid dissemination of research output.

The series covers the theory, applications, and perspectives on the state of the art and future developments relevant to systems and networks, decision making, control, complex processes and related areas, as embedded in the fields of interdisciplinary and applied sciences, engineering, computer science, physics, economics, social, and life sciences, as well as the paradigms and methodologies behind them.

Indexed by SCOPUS, EI Compendex, INSPEC, WTI Frankfurt eG, zbMATH, SCImago.

All books published in the series are submitted for consideration in Web of Science.

For proposals from Asia please contact Aninda Bose (aninda.bose@springer.com).

Grzegorz Sierpiński · Sebastian Naumann ·
Elżbieta Macioszek
Editors

Transport Systems and Urban Logistics

Springer

Editors
Grzegorz Sierpiński
Silesian University of Technology
Gliwice, Poland

Elżbieta Macioszek
Silesian University of Technology
Gliwice, Poland

Sebastian Naumann
Dipl.-Inf.
Institut f. Automation und Kommunikation e.V
Magdeburg, Germany

ISSN 2367-3370 ISSN 2367-3389 (electronic)
Lecture Notes in Networks and Systems
ISBN 978-3-031-81776-2 ISBN 978-3-031-81777-9 (eBook)
https://doi.org/10.1007/978-3-031-81777-9

This Springer imprint is published by the registered company Springer Nature Switzerland AG
The registered company address is: Gewerbestrasse 11, 6330 Cham, Switzerland

Preface

Sustainable development should be based on identification of needs, take the limited natural resources into account, and initiate technological development. The growth of city economy is strongly linked with how transport functions. The choices made by the travelling population, pertaining to how one travels, by what means of transport, by which transfer routes etc., globally translate into the negative impact of transport which takes diverse forms, primarily that of the emission of harmful substances, noise and the resulting traffic obstructions caused by congestion. In line with the European Union guidelines, minimisation of the negative environmental impact of transport cannot be pursued through mobility limiting, but rather through efficient utilisation of natural resources. The task of improving transport systems towards sustainable mobility requires adequate information transfer as well as systems capable of supporting integration of operations.

This publication contains selected papers submitted to and presented at the 20TH "Transport Systems. Theory and Practice" Scientific and Technical Conference organised by the Department of Transport Systems, Traffic Engineering and Logistics at the Faculty of Transport and Aviation Engineering of the Silesian University of Technology (Katowice, Poland). The problems addressed in the publication entitled *Transport Systems and Urban Logistics* include results of case study-based scientific and practical approaches to the problems of contemporary transport and logistics systems.

The authors drew attention among others to optimal realization of supply chains, especially on the last-mile stage, problems regarding to development of electric cars and, more broadly, the planning of charging infrastructure, traffic organization on intersections and parking locations, challenges for public transport including infrastructure condition, accessibility and fare strategies The publication also includes an abundance of numerical analyses based on significant data sets, illustrating the close affiliation between mobility and quality of life.

We would like to express our deepest gratitude to all authors, for reflecting the key problems of contemporary transport systems in a concise manner, as well as to reviewers, in recognition of their insightful remarks and suggestions without which this collection of papers would have never been published.

Katowice, Poland
September 2024

Grzegorz Sierpiński
Sebastian Naumann
Elżbieta Macioszek

Organization

20th Scientific and Technical Conference "Transport Systems. Theory and Practice" (TSTP2024) is organized by the Department of Transport Systems, Traffic Engineering and Logistics, Faculty of Transport and Aviation Engineering, Silesian University of Technology (Poland).

Organizing Committee

Organizing Chair:	Grzegorz Sierpiński	
	Silesian University of Technology, Poland	
Members:		
Agnieszka Popowicz	Ireneusz Celiński	Adrian Barchański
Grzegorz Karoń	Marcin Kłos	Iga Przytuła
Aleksander Sobota	Piotr Soczówka	Krystian Szewczyński
Marcin Staniek		
Renata Żochowska		

The Conference took place under the Honorary Patronage

Ministry of Infrastructure
Marshal of the Silesian Voivodeship

Scientific Committee

Co-Chair: Grzegorz Sierpiński	Silesian University of Technology, Poland
Co-Chair: Elżbieta Macioszek	Silesian University of Technology, Poland
Rahmi Akçelik	SIDRA SOLUTIONS, Australia
Werner Brilon	Ruhr-University Bochum, Germany
Jacek Chmielewski	Cracow University of Technology, Poland
Margarida Coelho	University of Aveiro, Portugal
Csaba Csiszár	Budapest University of Technology and Economics, Hungary
Roberta Di Pace	University of Salerno, Fisciano, Italy
Mehmet Dikmen	Baskent University, Turkey
Grzegorz Dydkowski	University of Economics in Katowice, Poland
Domokos Esztergár-Kiss	Budapest University of Technology and Economics, Hungary
Zoltán Fazekas	Institute for Computer Science and Control, Hungary
Piotr Folęga	Silesian University of Technology, Poland
József Gál	University of Szeged, Hungary

(continued)

Anna Granà	University of Palermo, Italy
Andrzej S. Grzelakowski	Gdynia Maritime University, Poland
Mehmet Serdar Güzel	Ankara University, Turkey
Józef Hansel	AGH University of Science and Technology, Cracow, Poland
Libor Ižvolt	University of Zilina, Slovakia
Marianna Jacyna	Warsaw University of Technology, Poland
Ilona Jacyna-Gołda	Warsaw University of Technology, Poland
Kazimierz Jamroz	Gdańsk University of Technology, Poland
Nan Kang	Tokyo University of Science, Japan
Jan Kempa	Bydgoszcz University of Science and Technology, Poland
Michael Koniordos	Pireaus University of Applied Sciences, Greece
Stanisław Krawiec	Silesian University of Technology, Poland
Andrzej Lewiński	Kazimierz Pulaski University of Technology and Humanities in Radom
Bogusław Łazarz	Silesian University of Technology, Poland
Zbigniew Łukasik	Kazimierz Pulaski University of Technology and Humanities in Radom
Michal Maciejewski	Technical University Berlin, Germany
Ján Mandula	Technical University of Košice, Slovakia
Antonio Masegosa	IKERBASQUE Research Fellow at University of Deusto Bilbao, Spain
Agnieszka Merkisz-Guranowska	Poznań University of Technology, Poland
Anna Mężyk	Kazimierz Pulaski University of Technology and Humanities in Radom
Leszek Mindur	International School of Logistic and Transport in Wrocław, Poland
Maciej Mindur	Lublin University of Technology, Poland
Goran Mladenović	University of Belgrade, Serbia
Kai Nagel	Technical University Berlin, Germany
Piotr Niedzielski	University of Szczecin, Poland
Przemysław Niewiadomski	University of Zielona Góra, Poland
Piotr Olszewski	Warsaw University of Technology, Poland
Enrique Onieva	Deusto Institute of Technology University of Deusto Bilbao, Spain
Asier Perallos	Deusto Institute of Technology University of Deusto Bilbao, Spain
Hrvoje Pilko	P3Mobility Ltd
Antonio Pratelli	University of Pisa, Italy
Dariusz Pyza	Warsaw University of Technology, Poland
Cesar Queiroz	World Bank Consultant (Former World Bank Highways Adviser), Washington, DC, USA
Andrzej Rudnicki	Cracow University of Technology, Poland
František Schlosser	University of Žilina, Slovakia
Ewelina Sendek-Matysiak	Kielce University of Technology, Poland
Jacek Skorupski	Warsaw University of Technology, Poland

(continued)

Aleksander Sładkowski	Silesian University of Technology, Poland
Wiesław Starowicz	Cracow University of Technology, Poland
Andrzej Szarata	Cracow University of Technology, Poland
Antoni Szydło	Wrocław University of Technology, Poland
Grzegorz Ślaski	Poznań University of Technology, Poland
Andrew P. Tarko	Purdue University West Lafayette, USA
Agnieszka Tubis	Wroclaw University of Technology, Poland
Frane Urem	Polytechnic of Šibenik, Croatia
Hua-lan Wang	School of Traffic and Transportation, Lanzhou Jiaotong University, Lanzhou, China
Mariusz Wasiak	Warsaw University of Technology, Poland
Adam Weintrit	Gdynia Maritime University, Poland
Andrzej Więckowski	AGH University of Science and Technology, Cracow, Poland
Adam Wolski	Polish Naval Academy, Gdynia, Poland
Olgierd Wyszomirski	University of Gdańsk, Poland
Elżbieta Załoga	University of Szczecin, Poland
Stanisława Zamkowska	Blessed Priest Władysław Findysz University of Podkarpacie in Jaslo
Jacek Żak	Poznań University of Technology
Jolanta Żak	Warsaw University of Technology, Poland
Renata Żochowska	Silesian University of Technology, Poland

Referees

Piotr Czech	Elżbieta Macioszek
Roberta Di Pace	Grzegorz Sierpiński
Marcin J. Kłos	Grzegorz Wojnar

Contents

Smart City Tools Supporting Sustainable Transport in Wrocław

Veronika Yavorskaya[1], Mateusz Rydlewski[1] (✉), and Emilia Skupień[1]

[1] Wroclaw University of Science and Technology, Wrocław, Poland
258892@student.pwr.edu.pl; mateusz.rydlewski@pwr.edu.pl; emilia.skupien@pwr.edu.pl

Abstract. The article discusses the issue of sustainable transport in cities, supported by Smart City tools. The authors present an overview of tools influencing the functioning of transport systems in a city, such as Intelligent Traffic Management Systems (ITMS), Mobility as a Service (MaaS), and Autonomous vehicles, and their role in creating a sustainable transport system. The Smart City tools implemented in Wrocław's (Poland) transport systems (ITMS and MaaS) were identified and analysed. Then, based on the described solutions, elements that could improve the effectiveness and efficiency of Wrocław's transport systems were identified and recommendations were stated. Since Wrocław uses ITMS and MaaS, the recommendations concern the development of these technologies and the introduction of AV into public transport in the city.

Keywords: Smart city · Transport systems · Sustainable transport · MaaS · Traffic systems · Autonomous vehicles

1 Introduction

The world has more and more inhabitants, and at the same time, an increasing percentage of them live in cities and industrialized areas [1]. For this reason, cities are growing and their functionality is becoming more important in order to properly serve their residents. The city should fulfil a number of functions that support both its inhabitants and its role in the broader regional, national, and global context. The city's key functions are:

1) Economic function - carried out by creating and maintaining jobs in various sectors of the economy; entrepreneurship support; support for trade and services.
2) Administrative function, including: management; security, and utilities.
3) Residential function - The city should offer a variety of housing options avail-able to all social groups, within this function it is also important to meet basic needs through appropriate water and sewage infrastructure, energy, and access to clean water.

G. Sierpiński et al. (Eds.): TSTP 2024, LNNS 1257, pp. 1–19, 2024.
https://doi.org/10.1007/978-3-031-81777-9_1

4) Social function, including education; Healthcare; culture, recreation, integration, and social participation.
5) The function of caring for natural resources, i.e., protecting the environment; care for urban greenery and sustainable development.
6) Transport function - Implemented through effective and accessible public transport; road infrastructure; and broadly understood communication [2].

Due to technological development, tools of this field are also involved in the city's fulfilment of its functions. Supporting the functioning of the city through information and communication technologies is included in the Smart City concept. This term has many definitions, but all of them refer to the city's development strategy, focussing on creativity, innovation, and flexibility, understood as the ability to quickly adapt to external and internal conditions, while respecting sustainable development. The Technical University in Vienna identified six key dimensions of Smart Cities: Smart Economy, Smart Environment, Smart People, Smart Living, Smart Governance and Smart Mobility [3]. The main goal of Smart City is to improve the standard of living of residents and increase their participation in making important decisions. And one of the key elements is transport, which has a huge impact on the implementation of all other functions of the city.

Transport systems and the quality of the services they offer influence the eco-nomic function of the city. Of course, some trade and services take place online, but a significant part of the service to residents and the industrial sector of the city requires transport of goods. Additionally, it is necessary to move people related to the functioning of this area of the economy.

Administration is an area that increasingly uses the Internet of Things, so we deal with more and more so-called administrative matters online, but some of them still require personal presence at the office, and city administration requires officials to know its topography and local problems, which again works well with the need of moving around.

The residential function of the city is directly connected with the transport function, due to the attractiveness of the location related to the possibility of accessing key traffic attractors and the movement of residents as such. Additionally, residential estates must also be supplied with goods (shops, personal deliveries) and municipal utilities.

All social functions assume social interactions, which, with a few exceptions, are face-to-face, so they require the movement of people to facilities where these functions can be implemented.

Care for natural resources, implemented through transport, focusses primarily on reducing the impact of transport on the natural environment (through ecological sources of power for vehicles, appropriate infrastructure, and organising space in a way that is friendly to pedestrian and bicycle travel).

Furthermore, appropriate control of transport systems allows increasing the system's resilience by significantly impacting congestion, efficiency, comfort, safety, flexibility, ecology, and economy [4]. All this significantly affects the quality of life of city residents.

The concept of sustainable transport is related to the definition of sustainable development, which should meet current needs without compromising the ability of future generations to meet them. According to Richardson [5], in a sustainable transport system, the impact on the natural environment (fuel consumption, vehicle emissions) and the impact on the quality of life of residents (congestion, safety and socioeconomic accessibility) should be maintained at a level that does not cause harm to subsequent generations.

However, some researchers often focus only on environmental issues (resource depletion or air pollution). Although this problem represents the most significant long-term threat, it is inappropriate to equate sustainable transport solely with environmental aspects. It is important to also analyse social and economic well-being. Sustainable transport includes integrating environmental aspects with social and economic issues, but also building an appropriate institutional environment for this purpose [6].

The aim of this article is to present an overview of the tools that influence the functioning of transport systems in a city. The tools of Smart City that could improve the sustainability of Wrocław's transport systems are identified and recommendations are stated.

2 Tools Supporting Transport Systems

Traffic in city residents is an integral part of their life in metropolis. Many needs can be identified in urban areas, the most important of which include the following needs: food, education, housing, services, culture, health, recreation and sports, living, working, social needs, and public administration [7]. The transport need is the entity that connects the possibility of providing each of these services. This can be implemented in many ways, but in each case it must perform the transport function between the source and the destination. This need can be carried out individually, by selecting the means of transport by the person making the trip, e.g., a bicycle, scooter, car or other means of transport, either owned or rented (as part of the sharing option). Another form is to travel using public transport or other external transport services. These journeys can also be carried out without using additional means of transport, such as walking journeys, which, as nonrecreational journeys, are most often carried out over short distances.

Bearing in mind that modern cities are characterised by the need to meet many of these needs, it is considered a very important goal of a Smart City to provide a reliably built environment for citizens (buildings, transport, and urban infrastructures). The purpose of these activities is (a) an increase in productivity and competitiveness; (b) a reduction in the environmental impact; and (c) a better welfare for its citizens [8].

As part of the broadly understood area of transport systems in the Smart City concept, the authors identified, among others: Intelligent traffic management systems, Mobility as a service (MaaS), Autonomous vehicles, Public transport systems (including technologies supporting fleet management), Sustainability transport (including shared mobility, walking and cycling mobility), Energy efficiency and ecology, Data management and big data analysis.

Due to the breadth of the identified topics, the first three issues are discussed in detail in the following chapter. Each of them was supported by information about the systems operating in certain areas.

2.1 Intelligent Traffic Management Systems (ITMS)

Intelligent Traffic Management Systems (ITMS) are advanced technological solutions that utilise sensors, computers, electronics, and communication technologies to enhance the safety and efficiency of transportation systems. ITMS are crucial components of Smart Cities, integrating traffic management, vehicle monitoring, and passenger information systems. The role of ITMS is important due to its effects including optimising traffic flow, reducing travel time, and minimising CO_2 emissions, contributing to improved urban living standards and sustainable urban development [9].

Vehicle-to-Everything (V2X) communication which includes Vehicle-to-Vehicle (V2V), Vehicle-to-Infrastructure (V2I), Vehicle-to-Network (V2N) and Vehicle-to-Pedestrian (V2P) is integral to the development of ITMS and plays a crucial role in achieving the goals of Smart Cities by enhancing mobility, safety, and sustainability. V2V communication enables direct data exchange between vehicles to enhance road safety and traffic efficiency. This technology allows vehicles to share real-time information such as speed, position, direction, aiding in collision avoidance, and coordinated driving. On the other hand, V2I communication involves interaction between vehicles and road infrastructure such as traffic signals and roadside units (RSUs). This technology provides drivers with real-time updates on road conditions, traffic signals, and potential hazards. Studies [10] have shown that cellular networks, particularly 5G, offer significant advantages in delivering timely and accurate information, enhancing overall traffic management and safety. The V2P system enhances pedestrian safety by enabling vehicles to detect and communicate with pedestrians carrying connected devices such as smartphones. V2P systems can alert drivers to the presence of pedestrians, particularly in situations where visibility is limited, which can help prevent accidents. V2N communication refers to the connection between vehicles and broader network infrastructures, such as cellular networks. This type of communication supports a range of services, including traffic updates and navigation assistance. V2N facilitates better traffic management and enhances the overall driving experience by ensuring that vehicles have access to the necessary data from network services [10].

2.1.1 Challenges of Intelligent Traffic Management Systems (ITMS). Due to the fact that the number of vehicles on the roads has been constantly growing [11], the concept of ITMS is fundamental to solve the problems of transportation. By integrating advanced technologies such as machine learning, artificial intelligence, and wireless communications, ITMS promises to enhance traffic management, improve safety, and reduce environmental impact. However, like any groundbreaking innovation, the widespread adoption of ITMS is not without challenges covering regulatory and standardisation issues, technological advancements, and societal acceptance. Addressing these obstacles is crucial to achieving the full potential of ITMS in creating smarter, more efficient, and sustainable transportation systems.

Ensuring the interoperability of ITMS technologies across different regions and platforms requires uniform regulatory frameworks and standards. The absence of such standards can impede the integration and effectiveness of ITMS solutions, posing significant barriers to their deployment and scalability. Clear regulations are needed to define legal responsibilities and liabilities in the event of accidents or malfunctions involving ITMS. Without such frameworks, it can be challenging to determine accountability, which can lead to legal disputes and hinder public trust in ITMS technologies.

The lack of public trust in Intelligent Traffic Management Systems is a significant barrier to their successful implementation and widespread adoption. This scepticism can arise from several factors, including concerns about privacy, security, reliability, and the overall impact on daily life. Understanding these concerns is essential to address them and foster a supportive environment for ITMS deployment. Ethical concerns, such as the fairness of algorithms and the potential for bias in decision-making processes, can also affect trust. Ensuring that ITMS are developed and implemented in a socially responsible manner is crucial to gaining public acceptance.

Intelligent Traffic Management Systems are designed to enhance the efficiency and safety of transportation networks by enabling vehicles to communicate with each other and with road infrastructure, but increased connectivity and reliance on real-time data exchange introduce significant cybersecurity risks. Insecure communication channels can be exploited to inject false information, leading to traffic disruptions and accidents. In the article [12] authors emphasize the importance of ensuring digital security for the development of ITMS technology. ITMS, especially vehicular ad hoc networks (VANETs), are highly susceptible to cyberattacks due to their interconnected nature. These attacks can disrupt traffic management systems, compromise vehicle safety, and cause significant economic losses. On the other hand, the extensive data collection by ITMS, including vehicle locations, travel patterns, and personal information, raises privacy concerns. Unauthorised access to these data can lead to privacy breaches and misuse of personal information. The article [12] emphasizes the importance of a thorough risk analysis to identify and mitigate potential threats.

Intelligent Traffic Management Systems are the cornerstone of Smart City infrastructure, leveraging advanced technologies to create efficient, safe, and sustainable urban environments. ITMS integrate real-time data collection, analysis, and communication to enhance urban mobility, reduce environmental impact, and improve overall quality of life. First of all, ITMS improve the safety of the city and supports public transport optimization. Real-time passenger information systems provide updates on transit schedules, delays, and alternative routes, which improve the user experience and encourage the use of public transport. This reduces the reliance on private vehicles, alleviating traffic congestion and lowering emissions [11].

2.1.2 ITMS solutions based on an example from Wroclaw (Poland). Implementing ITMS helps to make decisions about urban planning by delivering huge amounts of data. These data help city planners understand traffic patterns, identify bottlenecks, and develop strategies to improve transportation networks. Data analysis and machine learning algorithms are used to predict traffic trends and optimise urban mobility solutions. The deployment of autonomous vehicles is heavily dependent on ITMS infrastructure. Autonomous vehicles rely on ITMS for navigation, traffic management, and safety features, which are essential for their safe and efficient operation in urban environments.

ITMS has been operating in Wrocław since 2012 and its main goals at the time of implementation were: improving the traffic of public transport vehicles and cars in the city and increasing road traffic safety. At the time of its implementation, this system was the largest one-off implemented system of this type in the country. Its operation is achieved, among other things, by controlling traffic lights in real time, making it possible to adjust the length of phases and their sequence to traffic conditions. The vehicle detection system is based on three pillars:

- Vision detection (camera) – mainly used to identify and categorise vehicles in lanes in the area of intersections,
- Inductive detection (induction loops on the surface, tracks, or traction network) – used both for the detection of trams (and buses) and cars,
- Radio detection – used only to detect public transport vehicles.

Additionally, as part of the existing vehicle-system communication channels in the tram system in Wrocław, there is also infrared communication. This type of information transmission is used by tram drivers to include information about the direction of travel and the appropriate setting of the semiautomatic track switch.

The vision system in Wrocław includes three types of cameras [13]: ARTR cameras and two types of velocity cameras. ARTR cameras have functionality that includes the ability to read licence plates. These data are sent to the system, and, based on the actual driving of vehicles around the city, the system is able to return information about how long it takes to travel on a specific road section. These data appear, among others, on gantries (large electronic boards with variable content), which inform vehicle drivers about the current situation on the road. On the one hand, it may be mentioned by travelling to a specific point (e.g. distinguishing different travel routes). Other information may include announcements about renovation works, street closures, or emergencies that are important from the drivers' point of view. Velocity cameras of the first type constitute the largest share of all cameras included in the ITMS system in Wrocław. They allow one to determine the vehicle's speed and length. Using the implemented software, they make it possible to distinguish seven types of vehicle, including passenger cars, buses, and delivery vehicles. These data are particularly important, for example, in determining the generic structure of traffic on a given road section or intersection. The last type of cameras used are second-type velocity cameras. These cameras record the number of vehicles and their mean speed in 10-min intervals. They are used on road sections and most often register vehicles in two opposite directions at the same time.

The functionality of the ITMS system was established at three levels: strategic, territorial and local [14]. The strategic level includes defining and planning strategies with the possibility of their modification, ensuring the possibility of coordinating the work of the entire system in various areas, supporting alternative routes for various forms of transport, and providing real-time measurement methods that allow assessing the effectiveness of the adopted actions and solutions. The territorial level includes measuring devices and detectors that allow the collection of data necessary to assess the quality of control. The local level includes activities related to accommodative control of traffic

Table 1. Benefits from the implementation of the ITMS system in Wroclaw (Poland), source [16].

Benefits from the implementation of the ITMS system	BEFORE	AFTER
Reduce time changes at the crossroads	2 months	1 day
Coordination of work lights for trams	Random, fixed-length cycles	Adapted to the situation at the crossroads
Flow of information e.g. on difficulties	Difficult	Immediately
Information on travel times of public transport vehicles	Lack	Shown in real time
The possibility of reaction	After departure and observations in the place of	Remote
Improve traffic measurement	Once in several years manually	Permanent and continuous measurement
Improving the safety of passengers and drivers	According to the procedures,	Integration of services, shortened response time special events
Support services, e.g. the police	Impossible	Create an electronic archive of video and archiving the state of traffic lights

lights, taking into account notifications and priority provision for public transport vehicles.

The ITMS system in Wrocław is additionally equipped with elements that serve primarily passengers of public transport and drivers of individual vehicles. The ITMS system in Wrocław collects and processes data in real-time, thanks to which it is possible to obtain information about the location of each tram and bus running in the public transport system in the city. These data can be used both in the context of providing passenger information at stops (dynamic passenger information boards) and also in the field of trip planning and verifying the location of vehicles individually by travellers.

Data from Wrocław ITMS are used, among others, to estimate travel times using neural networks [15]. The implementation of this system brings numerous benefits; in addition to reducing time (both for people travelling around the city and the time needed to obtain traffic data), the system also provides access to tools supporting traffic maintenance in the city. In [16], the benefits of implementing the system were collected and presented in Table 1.

2.2 Autonomous Vehicles

Autonomous vehicles (AVs) are poised to revolutionise urban transportation systems by taking advantage of cutting-edge technologies such as artificial intelligence (AI), machine learning, and advanced sensors. These vehicles operate without human intervention, making decisions based on real-time data from their environment. They are a critical component of smart city infrastructures, promising to enhance safety, improve traffic efficiency, and reduce environmental impacts. Autonomous vehicles significantly improve traffic management by optimising traffic flow and reducing congestion by communicating with traffic signals and other vehicles to coordinate their movements.

AVs rely on a combination of LiDAR, radar, cameras, and ultrasonic sensors to perceive their surroundings. These sensors collect data that is then processed by AI and machine learning algorithms to make driving decisions, such as lane keeping, obstacle avoidance, and traffic sign recognition [17]. One of the primary benefits of AVs is their potential to improve road safety. Human error is a major cause of traffic accidents and AVs can mitigate this by consistently applying safe driving practices. They are equipped with Advanced Driver Assistance Systems (ADAS) that include features such as automatic braking, lane keeping assistance, and collision avoidance. These technologies significantly reduce the likelihood of accidents, saving lives and reducing injury rates.

The advantage of AVs is that they provide enhanced mobility solutions for people who cannot drive, such as the elderly, disabled, and those without a driver's licence. By offering reliable and convenient transportation options, AVs improve the quality of life of these individuals and increase their independence. They also have a positive influence on quality of life by reducing emissions by optimising fuel consumption. Efficient driving patterns, such as maintaining steady speeds and minimising unnecessary braking and acceleration, help reduce fuel use. Furthermore, the integration of electric autonomous vehicles (EVs) further reduces the carbon footprint, contributing to cleaner and greener urban environments [18].

Shared Autonomous Vehicles (SAVs), also referred to as autonomous taxis, are a promising innovation in transportation that combines the concepts of autonomous driving and shared mobility. SAVs are designed to reduce the need for personal car ownership by offering on-demand transportation services. These vehicles can be hailed by users as needed, providing a flexible and convenient alternative to private car use. The primary objective of SAVs is to improve the efficiency of urban transportation systems, reduce traffic congestion, and reduce environmental impacts by optimising vehicle use and reducing the total number of vehicles on the road [19].

2.2.1 Challenges of Autonomous Vehicles (AVs). The appearance of autonomous vehicles (AVs) represents a monumental shift in the landscape of transportation technology. As these self-driving vehicles move closer to becoming a staple on our roads, a multitude of challenges and issues must be addressed to ensure their successful integration into society. The technological advancements that have propelled AV development forward are met with equally significant hurdles in areas such as regulatory compliance, safety, infrastructure, and public acceptance [20].

Figure 1 illustrates the complex nature of autonomous car development and the various factors that must be considered. The technical challenges include issues related to sensor technology, machine learning, and artificial intelligence. This flow chart is a useful tool to understand and address the different challenges facing autonomous car development, and helps to identify the areas that need further research and development [20].

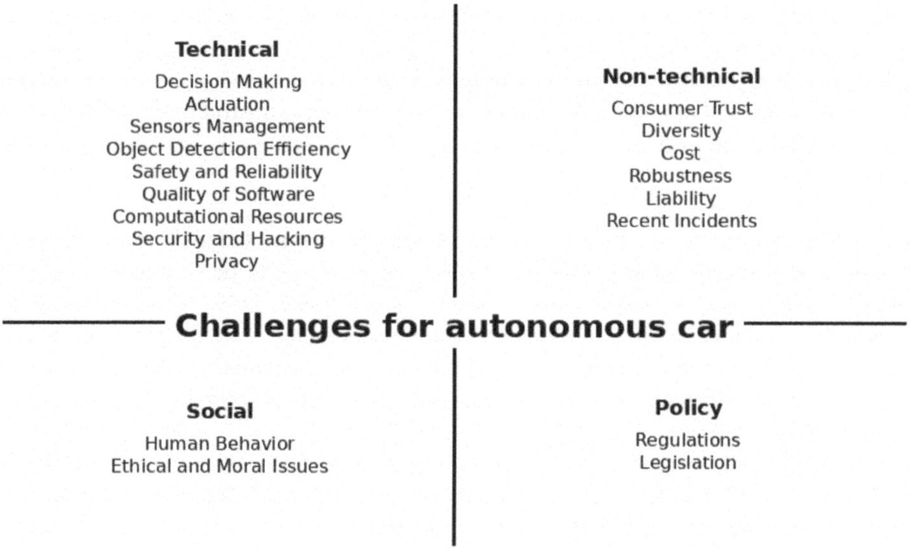

Fig. 1. Challenges for autonomous car. Based on [20].

Autonomous vehicles (AVs) face numerous technological challenges, including the reliability of sensors and the efficiency of machine learning algorithms used for perception and decision making. Real-time data processing and integration of various sensor inputs to create an accurate representation of vehicle surroundings are critical areas that require further development.

The validation and testing of AVs are complex and time-consuming processes that vary depending on the model's sophistication. Different testing techniques, ranging from simple bug fixes to comprehensive quality tests, are necessary to ensure that AVs meet rigorous safety and performance standards. Autonomous cars require extensive safety and mission-critical testing to fine-tune their performance and ensure all functionalities work as expected.

The software in AVs must handle a wide range of unanticipated situations while remaining fail-safe. Validating and testing these complex software systems is crucial to ensure that they respond appropriately under all conditions. AVs rely on numerous high-resolution cameras and sensors, such as LIDAR, for object recognition and environmental monitoring. These vehicles require powerful GPU processors with FPGA (field-programmable gate array) and SoC (system on a chip) to meet their functional and operational requirements. Connection to cloud infrastructure can provide the necessary computational power, but introduces potential issues with connection timeouts. Data shared between various vehicle components, other vehicles, and infrastructure must be protected from unauthorised access.

Public perception and acceptance of AV technology are crucial to its deployment. Addressing social impacts, including potential job losses in driving-related sectors, requires public policy and education. Building trust in AV technology through demonstrations of safety and reliability is essential to gain public support.

Integrating shared autonomous vehicles (SAVs) with public transportation systems poses operational challenges. Effective fleet management, coordination of passenger pick-ups and drop-offs, and ensuring seamless integration with existing public transport networks are areas that need to be addressed. In addition, infrastructure modifications, such as the creation of dedicated lanes or stops for SAVs, are required to support this integration [19].

2.2.2 AVs Solutions – Based on an Example from Gdańsk and Kraków (Poland). Shared autonomous vehicles (SAVs) can serve as feeder services, providing critical first-mile and last-mile connections to major transit hubs. This improves the accessibility and attractiveness of public transport, encouraging more people to use it, and thus reducing overall traffic congestion and emissions. For example, SAVs can transport passengers from residential areas to bus or train stations, effectively bridging the gap between public transport facilities.

Advanced traffic management systems can integrate SAVs with fixed-route transit services through dynamic traffic assignment models and multi-objective optimisation. These systems ensure that SAVs and public transport services operate efficiently together, optimising routes and schedules to meet demand and reduce passenger travel times.

Integrating AVs with public transport systems can have significant environmental and economic benefits. By reducing the need for private car ownership and optimising vehicle usage, cities can reduce emissions, decrease traffic congestion, and improve air quality.

As Smart City initiatives continue to evolve, Kraków is positioning itself as a leader in the integration of autonomous vehicle technology within its urban infrastructure. The city is embarking on various innovative projects aimed at harnessing AVs to improve public transport, reduce traffic congestion, and promote sustainable urban mobility. The aim of autonomous tram project [21] is to develop and implement an autonomous tram system to support tram drivers, improving their efficiency, and ensuring greater safety in public transportation. Initiated by the Institute of Rail Vehicles at the Faculty of Mechanical Engineering of the Kraków University of Technology, this research and development project focusses on creating an autonomous driving assistant for tram operators. The autonomous tram, known as "Nevelo," was produced by NEWAG SA from Nowy Sącz. The tram is equipped with a precise satellite navigation system and a control system that allows it to travel along the route without a driver. Key functionalities include speed monitoring, obstacle detection, automated braking and acceleration, and communication with traffic signals.

The autonomous tram project in Kraków represents a significant step forward in the city's Smart City strategy. By improving the integration of autonomous technology within public transport, this initiative is expected to deliver substantial environmental and economic benefits. Reduced dependence on private cars, lower emissions, and improved air quality are among the expected outcomes. Kraków plans to expand its autonomous vehicle initiatives further, future projects may include the deployment of autonomous trams on additional routes, integration with other forms of public transport, and collaboration with tech companies to continuously upgrade AV technology.

The city's proactive attitude towards embracing AV technology underscores its commitment to creating a more connected, efficient, and sustainable urban environment.

As part of the Sohjoa Baltic initiative [22], Gdańsk introduced autonomous electric minibuses to connect residents with major transit hubs. The pilot project initially ran from September 6 to October 4, 2019, where an autonomous minibus operated on a route from a parking area to the Gdańsk Zoo. The project continued in October 2021, with the bus now running within the Łostowicki Cemetery. The minibus travelled a 1.1 km route with four stops: Łostowicka, Sybiraków, Kolumbarium, and Cedrowa. This route was chosen to alleviate traffic congestion within the cemetery and provide transport for elderly and disabled visitors. Compared to other cities involved in the Sohjoa Baltic project, Gdańsk performed well in terms of user experience and acceptance of autonomous transport. Passengers in Gdańsk gave remarkably high scores for both personal security and safety on board the autonomous minibuses. The city received a higher overall experience score [23] compared to Helsinki and Kongsberg but was on par with Tallinn. The design and operational setup of the Gdańsk's route, which involved low-intensity environments, contributed to its positive reception. Gdańsk's autonomous vehicle projects represent significant advancements in the city's smart city strategy. By integrating AVs into public transport and logistics, these initiatives are expected to deliver substantial environmental and economic benefits. Reduced traffic congestion, lower emissions, and improved urban mobility are among the expected outcomes.

2.3 Mobility as a Service

Mobility as a Service (MaaS) is a transformative approach to urban transportation that integrates various forms of transport services, such as public transit, ride-sharing, car-sharing, bike-sharing, and other mobility options, into a single, cohesive, and user-centric digital platform. This integration allows users to plan, book and pay for multiple types of transport through one interface, enhancing the convenience, efficiency, and accessibility of urban mobility. MaaS aims to provide seamless, flexible, and sustainable transportation solutions that reduce the dependency on private car ownership, alleviate congestion, and reduce environmental impacts. The successful deployment of MaaS relies on advanced technological infrastructure, supportive regulatory frameworks, user acceptance, and robust data-sharing mechanisms to ensure smooth and efficient operation.

Mobility-as-a-Service plays a critical role in integrating various modes of urban transport, thus enhancing the efficiency and convenience of city transportation systems. By offering a seamless interface that combines public transit, ride-sharing, bike-sharing, and other mobility services, MaaS acts as a catalyst for urban transport integration. This integration is crucial to address urban mobility challenges such as congestion, pollution, and inefficiency. The user-centric design of MaaS platforms improves accessibility and convenience, making it easier for city dwellers to plan and execute their journeys using multiple transport modes [24].

MaaS platforms are designed to be user-centric, offering customised and flexible travel options tailored to individual needs. This approach increases convenience and satisfaction for users, encouraging the use of sustainable transport modes over private car ownership. The ability to plan, book and pay for journeys through a single application simplifies the travel experience, making public and shared transport more attractive [25].

Implementing MaaS can lead to significant environmental benefits by promoting the use of shared and public transportation over private vehicles. This change can reduce traffic congestion, reduce greenhouse gas emissions, and decrease the urban carbon footprint. MaaS encourages more sustainable travel behaviours, contributing to cleaner and healthier cities [26].

MaaS can optimise the use of existing transport infrastructure and services, leading to cost savings for both users and service providers. By improving the efficiency of resource allocation and reducing the need for new infrastructure investments, MaaS can lower operational costs and provide economic benefits to the city. Additionally, the pay-as-you-go model commonly used in MaaS platforms offers flexible pricing options that can be more cost effective for users [27].

MaaS platforms generate valuable data on travel patterns and user preferences, which can be used by city planners and policy makers to make informed decisions about transportation planning and infrastructure development. This data-driven approach allows for more responsive and adaptive transport systems that can better meet the needs of the city's residents [25].

2.3.1 Challenges and Issues in Mobility as a Service (MaaS). Transport plays an important role in the functioning of a city by ensuring the mobility of both people and goods. Transport integration can occur at multiple levels, including infrastructure, digital systems, information dissemination, tariffs, and institutional frameworks. MaaS can be considered as a final stage of transport integration in traditional terms [24].

One of the primary challenges in implementing MaaS is overcoming technological barriers. This includes the need for an advanced digital infrastructure to support real-time data sharing, integrated payment systems, and reliable connectivity across various transport modes. Ensuring interoperability between different technologies and platforms used by transport service providers is critical for the seamless operation of MaaS [24].

The implementation of MaaS is often hampered by regulatory and policy challenges. Existing transport regulations may not be flexible enough to accommodate the innovative models introduced by MaaS. Developing adaptive regulatory frameworks that can support the integration of various transport services while ensuring safety, privacy, and equity is crucial for successful MaaS deployment [25].

Another important aspect is user acceptance, which is a significant issue in the adoption of MaaS. Factors such as perceived convenience, cost, reliability, and trust in technology play crucial roles in influencing user behaviour. Addressing these concerns through effective communication, user-friendly interfaces, and reliable service provision is essential to encourage widespread adoption [26].

Ensuring that MaaS solutions are equitable and accessible to all users, including social groups that experience difficulties with handling new technologies or having access to banking, is a major challenge. Women, poorer, and less educated adults are more likely to be unbanked, meanwhile older age groups are not comfortable with using applications on smartphones. There is a risk that MaaS could exacerbate existing inequalities in urban mobility if not implemented inclusively. Policies and strategies must be developed to ensure that MaaS benefits all segments of the population [27].

Integrating various transport modes, including public transit, ride-sharing, car-sharing, bike-sharing, and maritime transport, into a single MaaS platform presents logistical and operational challenges [24], on the other hand, developing sustainable business models that ensure the financial viability of MaaS services is another significant challenge. The success of MaaS depends on finding a balance between providing affordable services to users and maintaining profitability for service providers [25].

MaaS plays a key role in the sustainable urban mobility planning (SUMP) framework, which integrates energy, transport, and information and communication technologies (ICT) to achieve urban sustainability goals. The European Commission emphasizes [28] the convergence of these three pillars to create efficient and sustainable urban mobility systems. By aligning MaaS with the goals of the 2030 Agenda for Sustainable Development [28], cities can enhance the quality of life of their residents through improved mobility solutions.

2.3.2 MaaS Solutions – Based on an Example from Wroclaw. A Smart City leverages data to make informed decisions about urban planning and management. MaaS platforms generate valuable data on travel patterns, user preferences, and system performance. These data can be used by city planners to optimise transport services, improve infrastructure planning, and improve overall urban mobility. The ability to analyse and act on real-time data helps cities become more responsive and adaptive to the needs of their residents. The adoption of MaaS in smart cities leverages technological advancements from the Fourth Industrial Revolution, such as sensors for the Internet of Things (IoT) and big data analytics, to create responsive and adaptive transport systems that meet the evolving needs of urban residents [29].

Urban mobility covers many different forms of transport, combining selected ones in intermodal journeys made by users to reach their destination efficiently, safely, comfortably, and quickly. The possibilities of combining individual car transport with public transport and/or the sharing option can be implemented, among others, through Park&Ride (P&R) parking lots. In [30] the importance of P&R parking lots in building and improving the quality of the public transport system in Wrocław was presented. In most cases, P&Rs operating in Wrocław are parking lots with limited access. This means that to enter them, one must have purchased a season ticket for public transport. Verification of possession of a ticket for a person driving a vehicle who wants to enter the P&R is carried out by scanning the QR code or swiping the city card on which the purchased tickets are encoded. This allows for quick access to available parking spaces but at the same time limits the availability of parking resources for people who want to potentially use them. This solution, as the author points out [30] *calls into question the principle of unlimited, universal, and free access and digitally excludes some users of the city, for example by a person interested only in using the car park.*

MaaS solutions that support the use of transport around the city include mobile applications that allow users to obtain a lot of information needed when implementing and planning a trip. It is also important that this information is provided reliably and in a clear way for the recipient. The public transport operator of the city of Wrocław allows its users to use the iMPK application. This application allows, among others, [3]:

to track public transport vehicles based on the GPS position, check whether they are going according to the schedule, view a photo of the type of bus that is going, and download the timetable from a specific stop to the phone. One can see if a given vehicle is late or is driving according to the schedule. There is also a timetable for passengers, along with information on the possibility of transfers. However, this application serves only information purposes. It does not have options that are important from the point of view of passengers planning to travel by public transport, i.e. the possibility of purchasing a ticket and planning a trip. The lack of this option reduces this application only to the information aspect and does not fit into the assumptions of MaaS, within which the integration of services is primarily implemented.

Wrocław, like many other cities in the country and around the world, provides data, including in real-time. The website dedicated to this is www.opendata.cui.wroclaw.pl, which contains data that can be divided into ten categories: spatial data, transport, city hall, other, environment, municipal waste, sports and recreation, education, culture, and social affairs. From the MaaS point of view, the most important data are gathered in the transport category, where one can find 13 different sets:

- Wrocław City Bike Price List,
- Number of vehicles moving around Wrocław during 1 day, divided into cities and provinces of Lower Silesia and other vehicles,
- Public transport schedule - current:
  - in .xml format,
  - in GTFS format,
- Wrocław City Bike rides - archived, for 2015 and 2016 (annual) and 2019–2021 (monthly),
- Faults of the Wrocław City Bike,
- Wrocław City Bike Stations - list of stations with geographical coordinates and the number of bikes available at the station,
- Sales of public transport tickets - quantitative and valuable data on ticket sales since 2016,
- List of bikes in Wrocławski Rower Miejski system,
- Wrocław City Bike rides,
- Parking lot occupancy - current reading - number of free spaces, number of vehicles entering and leaving, registration every 5 min in parking lots monitored in the ITMS system,
- Car parks filling – reading in 48 h,
- Location of public transport vehicles on the route - current location of buses and trams.

Wide access to data creates the potential and possibility of integrating this data in one place, the application. Unfortunately, currently there is no official city tool that would collect all this information in one place. Despite access to many of the data listed above, often constantly updated and providing current information, this potential is not used. From the point of view of a public transport passenger, two official city applications in Wrocław support aspects of mobility in the city: the iMPK application mentioned and described above and the Urbancard application, which gives the option to purchase a travel ticket and serves as a travel planner. The applications available on the market include many solutions that integrate data related to the current location of vehicles while providing the option of planning a trip and purchasing a ticket. These include, among others, the following applications: Jak-Dojadę, Google Maps and Moovit. However, each of them mainly refers to the public transport system.

On the other hand, there are many applications available on the European market that enable much greater integration of transport services in one place, expanding the options of the travel planner to search for alternative solutions, such as a taxi service, bike, or scooter rental. These applications often, in addition to the option of purchasing tram, bus, or metro tickets, also provide the opportunity to rent a city bike, or scooter, or order a taxi. This type of functionality is very important and convenient from the user's point of view. Examples of this type of solution in the world include the following applications: City mapper, transit, Jelbi, or BudapestGO. The Vooom application is available on the Polish market [31], the goals and described functionality of which correspond to the type that is important within the MaaS idea.

3 Smart City Tools Supporting Sustainable Transport in Wrocław

The solutions described in the paper are elements of the functioning of a Smart City. They constitute a very important basis, providing the potential to improve living conditions and the use of cities. The mobility of residents and their journeys should be supported by modern technologies that, thanks to their appropriate design, can serve city residents. The interplay between ITMS and MaaS solutions creates a major technological and regulatory challenge, especially in the context of the integration of private and public entities. The growing development of autonomous vehicle technology poses a significant challenge in the context of further integration of this system with already developed, often highly technologically advanced systems.

The basic benefits of the development of ITMS technology are as follows:

- Potential to reduce road congestion by improving traffic flow thanks to advanced traffic management technologies and real-time traffic light control.
- Increasing the attractiveness of public transport by improving the reliability of travel times, reducing the number of stops, and increasing commercial speeds.

The key benefits of developing autonomous vehicles include increased road safety by reducing human error and improved traffic flow due to optimised routing. Additionally, autonomous vehicles improve accessibility to transportation for the elderly and disabled, provide economic savings by reducing accident-related costs and operational inefficiencies, and help solve the problem of driver shortages.

The development of the MaaS concept creates the potential to increase:

- Convenience and accessibility to transport services for a wide range of recipients.
- Reduce the execution time of individual trips by planning solutions tailored to the user's needs in a wide range of transport options.

- Sustainable development by encouraging the use of other alternative means of transport while reducing dependence on private cars.

Wrocław, like many cities, is implementing ITMS and MaaS systems. Mobile applications for trip planning and ticket purchase, as well as public data in real time, are available to particularly facilitate travel and travel planning aspects. However, currently there is no single tool that integrates all of this information. This is an area for further improvement and development. The development of these systems can significantly improve the quality of urban transport, making it more efficient, attractive, sustainable, and accessible.

Taking into account the current state of ITSM in Wrocław and the capabilities of this system, the authors of this article recommend for Wrocław:

- improvement of the operation of Dynamic Passenger Information (DIP) systems and current information of residents about changes on these boards;
- optimisation activities to improve the throughput of intersections by modelling traffic in a computer environment based on current measurements of road traffic intensity, directional and generic structure. These activities would be carried out through corrections to traffic light programmes, or more precisely, the algorithms controlling them;
- using the system's capabilities to give real service priority to public transport vehicles;
- improvement of the detection system - earlier detection of vehicles or integration of the system with vehicle GPS (for public transport). This solution would be beneficial primarily for trams in the context of their detection before intersections on long routes (straight streets). Currently, they often have priority, but when reaching the intersection, they have to slow down, and with earlier detection, they would not have to. This would bring time benefits for the journey and environmental and economic benefits for the carrier (less energy consumption for braking and starting or accelerating);
- research into the use of bicycle detection (especially the possibility of detecting the direction of travel) in improving the smoothness of their movement through appropriate planning and organisation of traffic.

The recommendations for applications in the MaaS category are not specifically for Wrocław, but for cities in Poland in general. However, the operation of such solutions must be supported by the municipal services of each of the cities that are interested. The proposed directions of development are:

- real integration of functions: i) trip planning, ii) location and visualisation of stops (important for people with limited mobility) (preferably on different layers, enabling the inclusion of an appropriate level of detail), iii) location of city bikes, scooters, scooters, shared cars, free taxis;
- placing information about renovations, detours, route changes, breakdowns, timetables, vehicle locations divided into lines, stops, and individual cars (especially: the ability to filter low-floor cars);
- possibility of calculating for one route different variants for different vehicles (public transport, shared vehicles, taxis) travel time, costs, distance to be covered on foot;

- connecting all service providers from a given area (this will increase the competitiveness of entities and facilitate the conscious use of services by residents);
- possibility of purchasing tickets, paying for travel, or using a shared vehicle;
- in addition, as part of benefits for employees, companies could offer mobility packages and top up their accounts in such an application, e.g. at the level of 50% of the travel costs.

Since autonomous vehicles are still in the research phase, the recommendations for Wrocław do not assume implementation, but research, pilot implementation (in small areas/sections), and testing. As part of these activities, it is proposed to test the autonomous movement of trams at the depot (rearranging trains, lining up in the right order, working in conditions without people, automatically moving switches).

4 Conclusions

The basic goal of mobility in a city is to facilitate the traffic of individuals and goods, which can be brought through:

- reduction of the intensity of individual urban traffic,
- reduction of travel times,
- reduction of travel costs,
- reduction of environmental impacts.

These tasks are implemented through the Smart City concept and the sustainable development guidelines.

The solutions used in Wrocław described in the article fulfil their tasks, but they can be improved, especially MaaS, by integrating data from various sources and available in one application. The introduction of AVs to public transport services would help achieve better indicators for individual journeys of residents and take into account the difficulties on the driver market in Poland. ITMS operating in Wrocław operates satisfactorily and is constantly improved, affecting both travel times and, indirectly, also the impact of transport on the environment.

As stated in the introduction: sustainable transport should decrease its impact on the natural environment, but also increase its impact on the quality of life of residents. Despite the emphasis on digitalisation and the integration of smart technologies in the transportation sector, it is necessary to attach appropriate importance to the competency of officials and their levels of knowledge and skills, because sustainable development and high quality of life are largely a balance between the natural environment and technology.

The conclusions from the analysis presented in the article include further development of Intelligent Traffic Management Systems, increased participation of city authorities in the construction of an effective application integrating mobility data - Mobility as a Service, and continued research on autonomous vehicles that will be able to support the operation of public transport.

References

1. World Urbanization Prospects. The 2018 Revision, https://population.un.org/wup/Publications/Files/WUP2018-Report.pdf
2. Batty, M.: Cities as systems of networks and flows, vol. 14, 1st edn. Routledge (2017)
3. Ptak, A.: Smart city management in the context of electricity consumption savings. Energies. **14**(19), 6170 (2021). https://doi.org/10.3390/en14196170
4. Enjalbert, S., Ribeiro, S.K., Vanderhaegen, F.: Concepts and models about smart urban transport control systems for city resilience. IFAC-PapersOnLine. **53**, 10090 (2020). https://doi.org/10.1016/j.ifacol.2020.12.2732
5. Richardson, B.C.: Toward a policy on a sustainable transportation system. Transp. Res. Rec. **1670**(1), 27–34 (1999). https://doi.org/10.3141/1670-05
6. Roman, M.: Sustainable transport: a state-of-the-art literature review. Energies. **15**(23), 8997 (2022). https://doi.org/10.3390/en15238997
7. Szołtysek, J.: Podstawy logistyki miejskiej. Wydanie drugie (2009)
8. Ribeiro, P., Dias, G., Pereira, P.: Transport systems and mobility for smart cities. Appl. Syst. Innov. **4**(3), 61 (2021). https://doi.org/10.3390/asi4030061
9. Oladimeji, D., Gupta, K., Kose, N.A., Gundogan, K., Ge, L., Liang, F.: Smart transportation: an overview of technologies and applications. Sensors. **23**(8), 3880 (2023). https://doi.org/10.3390/s23083880
10. Tahir, M.N., Leviäkangas, P., Katz, M.: Connected vehicles: V2V and V2I road weather and traffic communication using cellular technologies. Sensors. **22**(3), 1142 (2022). https://doi.org/10.3390/s22031142
11. Guevara, L., Cheein, F.A.: The role of 5G technologies: challenges in smart cities and intelligent transportation systems. Sustain. For. **12**(16) (2020). https://doi.org/10.3390/su12166469
12. Zeddini, B., Maachaoui, M., Inedjaren, Y.: Security threats in intelligent transportation systems and their risk levels. Risks. **10**(5) (2022). https://doi.org/10.3390/risks10050091
13. Bazan, M., Bożek, M., Ciskowski, P., Halawa, K., Janiczek, T., Kozaczewski, P., Rusiecki, A.: Some aspects of intelligent transport system auditing. Arch. Transp. Syst. Telematics. **8**(3) (2015)
14. Trzcinowicz, B.: Rozwiązania dla transportu publicznego planowane w projekcie "ITS Wrocław". Przegląd Komunikacyjny (2011)
15. Ciskowski, P., Janik, A., Bazan, M., Halawa, K., Janiczek, T., Rusiecki, A.: Estimation of travel time in the city based on intelligent transportation system traffic data with the use of neural networks. Adv. Intell. Syst. Comput. **470** (2016). https://doi.org/10.1007/978-3-319-39639-2_8
16. Topolska, K., Topolski, T.: Capabilities for development an ITS system in the city of Wroclaw. Arch. Transp. Syst. Telematics. **8**(4) (2015)
17. Parekh, D., Poddar, N., Rajpurkar, A., Chahal, M., Kumar, N., Joshi, G.P., Cho, W.: A review on autonomous vehicles: progress, methods and challenges. Electronics. **11**(14) (2022). https://doi.org/10.3390/electronics11142162
18. Mo, T., Li, Y., Lau, K.T., Poon, C.K., Wu, Y., Luo, Y.: Trends and emerging technologies for the development of electric vehicles. Energies. **15**(17), 6271 (2022). https://doi.org/10.3390/en15176271
19. Carrese, F., Sportiello, S., Zhaksylykov, T., Colombaroni, C., Carrese, S., Papaveri, M., & Patella, S. M. (2023). The integration of shared autonomous vehicles in public transportation services: a systematic review. Sustain. For. (Vol. 15, Issue 17), 13023. doi:10.3390/su151713023

20. Padmaja, B., Moorthy, C.V.K.N.S.N., Venkateswarulu, N., Bala, M.M.: Exploration of issues, challenges and latest developments in autonomous cars. J. Big Data. **10**(1) (2023). https://doi.org/10.1186/s40537-023-00701-y
21. Górski, M., Konieczek, Z.: Autonomizacja jazdy tramwajem jako narzędzie wspierające pracę motorniczych. Transport Miejski i Regionalny. (2020)
22. FLOU ltd (Finland); Forum Virium Helsinki (Finland); City of Gdańsk (Poland); Chalmers University of Technology (Sweden); Finnish Transport and Communication Agency Traficom (Finland); IKEM (Germany): The Roadmap to Automated Electric Shuttles in Public Transport. User Experience and Impact on Public Transport (2020)
23. Bellone, M., Ismailogullari, A., Kantala, T., Mäkinen, S., Soe, R.-M., Kyyrö, M.Å.: A cross-country comparison of user experience of public autonomous transport. Eur. Transp. Res. Rev. **13**(1), 19 (2021). https://doi.org/10.1186/s12544-021-00477-3
24. Chmiel, B., Pawlowska, B., Szmelter-Jarosz, A.: Mobility-as-a-service as a catalyst for urban transport integration in conditions of uncertainty. Energies. **16**(4) (2023). https://doi.org/10.3390/en16041828
25. Maas, B.: Literature review of mobility as a service. Sustainability. **14**(14) (2022). https://doi.org/10.3390/su14148962
26. Zhang, Y., Kamargianni, M.: A review on the factors influencing the adoption of new mobility technologies and services: autonomous vehicle, drone, micromobility and mobility as a service. Transp. Rev. **43**(3) (2023). https://doi.org/10.1080/01441647.2022.2119297
27. Papaioannou, G., Polydoropoulou, A., Tsirimpa, A., Pagoni, I.: Assessing the potential of "mobility as a service" in passenger maritime transport. Front. Future Transp. **2** (2021). https://doi.org/10.3389/ffutr.2021.710311
28. Russo, F., Rindone, C.: Smart city for sustainable development: applied processes from SUMP to MaaS at European level. Appl. Sci. **13**(3) (2023). https://doi.org/10.3390/app13031773
29. Kim, S., Won, J., Kim, J., Lee, S.: The review on MaaS application methods for smart cities. Int. J. Transp. **6**(2) (2018). https://doi.org/10.14257/ijt.2018.6.2.04
30. Sokołowski, J.: Park&Ride car parks as an element of improving the quality of the transport system in Wroclaw. Biblioteka Regionalisty. **2022**(22), 105–115 (2022). https://doi.org/10.15611/br.2022.1.10
31. Jarosiński, K.: Historia rozwoju ruchu hulajnogowego wraz z charakterystyką wypożyczalni hulajnóg elektrycznych w Krakowie. Transport Miejski i Regionalny. (2021)

Choosing of Service Strategy for Cargo Owners in Regional Supply Chains

Oleksandra Orda[1](✉), Natalia Potaman[1], Grzegorz Sierpiński[2], and Viktor Bogomolov[1]

[1] Kharkiv National Automobile and Highway University, Faculty of Transport Systems, Kharkiv, Ukraine
orda@khadi.kharkov.ua

[2] Silesian University of Technology, Faculty of Transport and Aviation Engineering, Gliwice, Poland
grzegorz.sierpinski@polsl.pl

Abstract. Transportation is crucial in supporting production processes and distributing products. In this context, B2B logistics, which refers to the movement of goods from one business to another, is no longer a passive function but a transformative force driving business success. The proposed approach to choosing a strategy for servicing cargo owners assumes that regional delivery chains can utilize various alternative distribution channel structures. This approach considers the organization of cargo flows and the number of intermediaries (warehouses, distributors, retailers, etc.). The developed mathematical model allows for simultaneously considering a set of interrelationships of internal and external factors and choosing the best service strategy based on the efficiency criteria of minimum total costs. As a result of experimental studies the areas of effective use of alternative strategies and the structure of distribution channels by the requirements of cargo owners regarding the quality and cost of services were determined.

Keywords: Strategy · Distribution channel · Supply chain · Cargo owner · Regression model · Order volume · Delivery distance

1 Introduction

Under martial law in Ukraine, freight transportation has become a matter of national security for the smooth functioning of the economy, defense industry, population needs, etc. The efficiency and quality of transportation services play a crucial role in ensuring the stability and development of various sectors of the country, including business. Flexibility and reliability are modern requirements for efficient supply chains. Regional chains of B2B and B2C delivery of goods are sensitive to resource conservation, the energy crisis, the volatility of demand for products related to migration processes in the country, and risks caused by military operations. Third-party logistics (3PL) operators are essential for improving cargo services and ensuring efficient supply chains. Under the current conditions in Ukraine, storing large volumes of inventory at certain service areas to ensure consumer availability is unprofitable and risky for cargo owners

G. Sierpiński et al. (Eds.): TSTP 2024, LNNS 1257, pp. 21–32, 2024.
https://doi.org/10.1007/978-3-031-81777-9_2

(e.g., distributors). Therefore, transport support for production processes and product distribution constantly requires carriers and delivery operators to develop and adopt practical solutions to improve the efficiency of cargo owner services in supply chains to meet the needs of businesses and the country's population.

The gradual economic recovery and increased volume of road cargo transportation in Ukraine necessitate efficient and reliable supply chains that would ensure the sustainability and reliability of business processes. Focusing on excellent customer service can drive demand and help a business stay ahead of the competition [1]. When developing a service strategy, it is crucial to consider the distribution channel. The distribution channel can eliminate direct interaction with the end consumer and allow the manufacturer to reallocate resources from non-specialized delivery activities. Delivery operators responsible for various distribution channels often grapple with the vital question of which parameters to prioritize when structuring the distribution channel and choosing a strategy for servicing cargo owners. In practice, with time constraints on decision-making, delivery operators rely on previous service experience or make decisions intuitively when forming a distribution channel. This approach has negative consequences for business, as it doesn't consider the stochastic nature of the transportation service delivery process, the dynamism and the variety of types of interaction between participants in the "manufacturer-transporter-consumer" chain, etc. Therefore, to ensure the continuity of business processes, the issue of developing an approach to forming a rational distribution channel structure as a decision-support tool for delivery operators on the choice of strategies for servicing cargo owners in supply chains is relevant.

2 Literature Review

The future of freight transport in the EU will be defined by the integration of technologies into dynamic supply chains, with logistics centers playing a pivotal role [2]. According to the well-known concepts of supply chain management (Supply Chain Management, Collaborative Planning, Replenishment, and Forecasting, Efficient Consumer Response, etc.), alternative strategies for servicing cargo owners in supply chains should be formed to meet customer needs per the requirements by providing high-quality cargo delivery services at minimal cost based on the principles of organizing integration processes of interaction between participants [3–7].

Scientists pay considerable attention to the issue of choosing a strategy for servicing cargo owners, as it determines the commercial success of the entire supply chain. Most of the proposed approaches are aimed at solving the issues of routing and selection of rational rolling stock for operation on routes [3–5, 7], selection of a cargo delivery scheme [5–8], synchronization of demand [8], sales, distribution, and delivery parameters [8, 9], inventory management [1], resource-saving [5, 7, 8], strategic planning of mail and parcel distribution networks [9], etc. Various factors, including the cost of logistics services, the cost of sales, and the green production cost ratio, influence the choice of a service strategy. In the study [10], the authors investigate the impact of demand parameters, such as delivery distance and cargo weight, on choosing an optimal strategy for the freight forwarder's behavior in the transportation market. The proposed approach in the research [11] allows us to determine that under conditions of price competition, the optimal distribution policy between symmetric producers using two-channel supply chains is an asymmetric policy. The study [12] investigated how vertical integration decisions in logistics supply chains are increasingly driven by competition for service efficiency and how vertical integration of resources can im-prove service quality and demand, providing a competitive advantage.

Forming a distribution channel is a strategic decision that significantly impacts a business's short- and long-term success. Developing distribution channel structure in regional supply chains has attracted considerable attention in recent studies, especially with the development of new technologies in logistics, including digital ones, and in the context of sustainable development and greening of the transport sector following the European Green Deal, etc. As a tool, the authors in [13] use a multi-purpose model for optimizing a low-carbon logistics network as a vital component of the distribution of goods to minimize complex costs and transportation time based on a genetic algorithm. The structure of the product distribution channel in the supply chain is determined by and formed from the supply and commodity flow channels [3]. Control, as a management function, should be carried out in the distribution channel in each link from the supplier of raw materials to the end consumer. The results of the simulation of the interaction of product manufacturers, transport companies, and terminals presented in the research [14] can be used to guide decision-making in perishable product supply chains by identifying the most favorable conditions for different options subject to specific demand parameters. The paper [15] proposes an approach to designing a distribution network in a complex four-tier supply chain system using a mixed integer linear programming model. In addition to solving the tactical planning problem, the author analyzes the model's sensitivity to internal and external parameters: product demand, transportation cost per unit, and warehouse cost per unit, which allows for tactical management decisions in the distribution network during challenges. Researchers pay special attention to multi-echelon systems and their varieties. This article [16] presents the classification and existing approaches to solving a two-echelon issue, which is divided into three classes: a two-echelon location routing problem, a two-echelon vehicle routing problem, and a truck and trailer routing problem. The variety of types of interaction of participants in the supply chain necessitates its construction based on pooling resources, coordination of actions, creation of an integrated information space, optimization of costs for delivery and sale of products, etc. Therefore, transport companies should use a systematic approach when choosing strategies for servicing cargo owners. They should consider concepts such as material flow, distribution channel, and logistics system as one integrated system.

By summarizing the existing approaches, it is advisable to form a strategy considering such components as product, price, and communications. When addressing the issue of forming a rational structure of the product distribution channel in the supply chain, on which the efficiency of servicing cargo owners depends, much attention is paid to developing a marketing policy for product distribution. At the same time, in our opinion, insufficient attention has been paid to the study of the technological foundations of developing a distribution channel, which primarily depends on determining the set of fundamental technological and technical parameters of the delivery process that have the most significant impact on the structure of the distribution channel, especially in current conditions in the transport services market in Ukraine. Currently, the distribution system has many different distribution channels with other structures. All options for organizing the structure of the distribution channel have the right to exist. However, depending on the conditions of product distribution, manufacturers or intermediaries choose one or another distribution channel structure: direct delivery of products from the source of the cargo flow to the source of its absorption with the participation of mainline transport or with the involvement of a point (several points) of consolidation and distribution of the material flow through the participation of local and mainline transport.

This paper aims to develop an approach to improving the efficiency of servicing cargo owners in regional supply chains by choosing a service strategy based on forming a rational product distribution channel.

3 Research Methodology

In the distribution logistics field, choosing a distribution channel for material flow is a crucial management decision that significantly influences other choices related to flow process management. Given the many distribution channels available, creating an appropriate distribution channel from a set of alternatives requires high complexity. It is advisable to specify the number and types of channels under consideration in the study to facilitate the search for a viable solution within the system of constraints.

The research focused on the structural framework of cargo delivery systems along different variants of distribution channels (Fig. 1). The delivery system includes the main participants: consignor, consignee, carrier, and terminal. Each participant has an internal structure. The nature and properties of the object of study were revealed between participants and internal communication structures. During the analysis of the structure of the research object, it was determined that the delivery of consumer goods when servicing cargo owners in regional transportation could be realized using alternative variants for the structure of distribution channels, taking into account the schemes for organizing cargo flows and the number of intermediaries: direct option (the I Scheme), using points of distribution of material flows (the II Scheme), using points of consolidation and distribution of material flows (the III Scheme).

The structural outline of the functional links of the delivery system accounts for the following input data: the volume of customer order—g, tons; the number of consumers—n, units; the average distance of cargo delivery—Li, km; the time interval between requests for transportation services—I, hours; external factors—$\{Z\}$; characteristics of the alternative delivery schemes conditioned by assumptions and limitations.

The delivery process depends on the variants for the structure of distribution channels and consists of the following elements: accepting the request; loading of goods onto a vehicle at the consignor's warehouse; transportation of goods to the destination terminal (for the II Scheme and the III Scheme); unloading the car at the destination terminal (for the II Scheme and the III Scheme); storage of goods in the destination terminal (for the II Scheme and the III Scheme); transportation of goods to

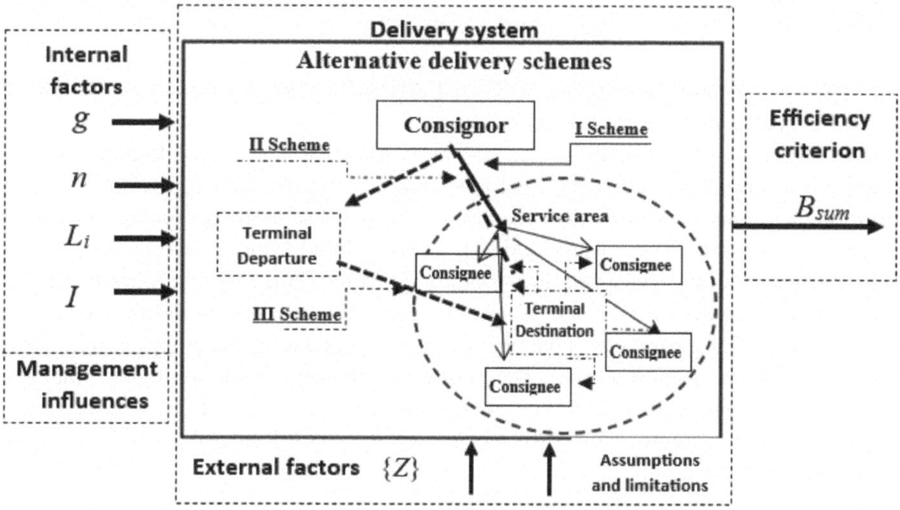

Fig. 1. Structural outline of the functional links of the delivery system.

the departure terminal (for the III Scheme); unloading the vehicle at the departure terminal (for the III Scheme); storage of goods in the departure terminal (for the III Scheme); loading of goods onto a vehicle at the terminals (for the II Scheme and the III Scheme); transportation of goods to the consumers; unloading at the consumer's warehouse; documents processing.

To choose a service strategy, it is essential to devise a rational distribution channel for transporting goods, prioritizing the total cost of transporting 1 ton of cargo as a critical efficiency criterion—*Bsum*, UAH/t. This emphasis is due to the substantial influence of this cost factor on the price of products for the final consumer. Reducing costs can be achieved by selecting an optimal structure for distributing consumer goods channels tailored to specific cargo transportation scenarios. Moreover, diminishing the overall cost of cargo delivery emerges as a critical strategy for reducing tariffs, which is vital in attracting additional clientele.

The research was based on the following accepted assumptions:

- product delivery is considered in long-distance transportation.
- The service area has the shape of a circle.
- The point of consolidation and distribution of material flows is located in the center of the territory to be served.
- Characteristics of freight vehicles are determined.
- The level of demand parameters are determined stochastically.

When delivery operators decide on a distribution channel, they aim to minimize costs. These costs, which encompass different aspects, all contribute to the total expenses that must be reduced. To gain a deeper understanding of the total costs, it is essential to identify the specific components involved in delivering 1 ton of cargo through the distribution channel. In this paper, the total costs encompass the following indicators:

$$B_{sum} = B_{Inv} + B_{TR} + B_{IS} \tag{1}$$

where *BInv*—inventory management costs, UAH/t; B_{TR}—transportation costs, UAH/t; B_{IS}—information support costs, UAH/t.

The operation of terminals drives inventory management costs. Warehousing of products at terminals is necessary due to the existing irregularities in production, transportation, and consumption cycles. Terminals can be set up at the beginning, middle, or end of transportation flows or production processes. In addition to warehousing operations, intra-warehouse transportation, loading and unloading, sorting, packaging, and transshipment are also performed. Therefore, terminal systems should be viewed not only as devices for storing cargo but also as transport and terminal complexes in which cargo flow plays an important role. The purpose of creating a functioning terminal system is not only to receive a cargo flow with specific parameters from transport, process it, and transfer it to another but also to process cargo with minimal costs. These costs also include insurance, taxes, rent for warehouse space if it does not belong to the system, etc.

Inventory management costs consist of cargo storage costs, cargo processing costs at the terminal or warehouse, and "freezing" turnover capital expenses. Inventory management costs can be fined by the following function:

$$B_{Inv} = t_n \cdot \left(\frac{1}{2} + \frac{R_c}{g \cdot n \cdot \gamma}\right) \cdot \left(1 + \frac{C_{prod.} \cdot H_d}{36500}\right) + \frac{S_{L-U} \cdot n \cdot g \cdot \gamma \cdot (t_{1L} + t_{1U}) + 2 \cdot C_m \cdot T_m \cdot \eta \cdot g \cdot n \cdot \gamma}{g \cdot n \cdot \gamma} \tag{2}$$

where *Rc*—the size of the insurance stock in the warehouse, tons; $S_{L\text{-}U}$—the cost of loading and unloading operations, UAH/h; $t_{1L}(t_{1U})$—the average time for loading (unloading) 1 ton of cargo, h; η—the portion of freight that is moved and sorted in the warehouse; *Cm*—the cost of maneuvering in the warehouse, UAH/h; *Tm*—the average for maneuvering in the warehouse, h; *Hd*—the discount rate, %; *Cprod*—the unit cost, UAH/t.

The leading logistics principles, typical for firms producing and consuming products (customer priority, high level of service, reduction of order fulfillment time, etc.), fully apply to the transport industry enterprises involved in logistics systems. It can be determined that a significant part of logistics operations on moving the material flow from the producer to the consumer is performed using various vehicles. Thus, transportation plays an essential role in promoting products and incurs certain costs. This approach considers the process of cargo delivery by road.

In setting the problem, it is assumed that consumer goods are delivered in long-distance traffic according to alternative variants of distribution channel structures, of which the II and III Schemes involve the use of distribution points or the combination of material flows (Terminals). Thus, the transportation of consumer goods is divided into two delivery components: long-distance delivery to distribution points or material flows consolidation and delivery in urban traffic from the points to the end consumers of consumer goods.

Long-distance transportation costs can be fined as follows:

$$B_{TR}^{IC} = \frac{\left[\frac{L_D}{\beta} \cdot \left(a_z + в_z \cdot g \cdot n \cdot \gamma + \frac{a_n + в_n \cdot g \cdot n \cdot \gamma}{V_T}\right)\right.}{g \cdot n \cdot \gamma} + \\ + \frac{\left.(a_n + в_n \cdot g \cdot n \cdot \gamma) \cdot (g \cdot \tau_{n(p)} \cdot n \cdot \gamma + t_g)\right] \cdot N_d}{g \cdot n \cdot \gamma} \quad (3)$$

where L_D—the average distance of cargo delivery, km; a_n—the minimum value at a constant cost component of road transportation, UAH/h; $в_n$—coefficient of change at a constant element in the cost of road transportation, UAH/h; a_z—the minimum value for the variable component of the cost of road transportation, UAH/km; $в_z$—the coefficient of change at the variable component of the cost of road transportation, UAH/t·km; V_T—the technical speed, km/h; β—a mileage utilization rate; $\tau_{n(p)}$—the time required for loading (unloading) of products, t/h; t_g—the additional time for processing documents, h; N—the number of end consignees, units; N_d—the number of vehicle trips, units.

Transportation costs within the city are calculated based on the assumption that the city's shape is assumed to be a circle, and the point of distribution or consolidation of the material flow is located in the center of the circle. In this case, the transportation costs are given as

$$B_{TR}^{WC} = \frac{\frac{2}{3 \cdot \beta} \cdot \sqrt{\frac{F_0}{\pi}} \cdot \left(a_z + в_z \cdot g \cdot N \cdot z + \frac{a_n + в_n \cdot g \cdot N \cdot z}{V_T}\right)}{g \cdot N \cdot \gamma} + \\ + \frac{(a_n + в_n \cdot g \cdot N \cdot z) \cdot (g \cdot \tau_{L(U)} \cdot N \cdot z + t_g)}{g \cdot N \cdot \gamma} \quad (4)$$

where F_0—the area of the region, km^2.

Information flows are the connecting threads that link all elements of the logistics system. Costs associated with information support are calculated by the following function:

$$B_S = \frac{S_1 \cdot I}{g \cdot n \cdot \gamma} \quad (5)$$

where S_1—the cost of processing one request, UAH/request.

Our hypothesis posits that a confluence of internal and external factors plays a pivotal role in determining an optimal strategy for servicing cargo owners in long-distance transportation. This strategy aims to minimize delivery costs while concurrently enhancing overall operational efficiency.

4 Results and Discussion

In analyzing the mathematical models based on formulas (1)–(5), a set of parameters has been identified that significantly impact the distribution channel's structure. These parameters were used in experimental studies to test the hypothesis. They include the following impact factors:

- the number of consumers of goods, units;
- the volume of consumer orders, tons;
- the size of the service area, km2;
- the average delivery distance in regional traffic, km;
- the time interval between requests in a flow, h.

The implementation of the proposed approach is presented as an example of the case of transport processes for the transport company Neolith Logistic (Ukraine), which organizes regional freight transportation and serves manufacturers of varnish decor and paints from Kharkiv. The company delivers products from manufacturers to wholesale (B2B) and retail (B2C) trade networks in Poltava, Kyiv, Dnipro, and Lviv.

To conduct modeling and obtain the most reliable results of experimental studies, the required sample size of observations was determined to analyze the technological parameters of the delivery process, in particular, technical speed, volume of customer orders, and interval of the request flow. The values of the variable parameters were obtained by processing orders for the delivery of products during 2023 in the total amount (general population) of 120 observations. Using the Statistica software, the characteristics of random variables of technological parameters and their distribution laws were determined. It is established that the values of the parameters of technical speed and volume of customer orders are distributed according to the normal distribution of random variables, and the parameter of the interval of the request flow is distributed according to the exponential law.

The range of variation of each exposure parameter involved in experimental studies is determined, and an experiment plan is developed based on the obtained statistical data (Table 1). The input parameters for the I Scheme of the delivery are four parameters, and for the other schemes are five parameters due to the alternative structures of the product distribution channel. The complete factorial experiment was conducted for the three proposed variants of the structure: for the I Scheme, 16 series of experiments were performed, and for the II and III schemes, 32 series of experiments were performed.

Table 1. Levels of variation of impact factors.

Factor	Value levels	
	Minimal	Maximal
Number of consumers of goods	20	200
Volume of consumer orders, t	0,5	10
Size of the service area, km^2	112	800
Average delivery distance in regional traffic, km	150	1016

To identify the dependencies between the variables, we conducted a regression analysis using linear and power models calculated using Microsoft Excel software. Based on the higher probability value of the power regression model obtained

Table 2. Regression models for alternative structures of distribution channel.

Structure of the distribution channel	Regression model
I Scheme	$B_{sum} = 2{,}557 \cdot g^{0{,}125} \cdot n^{0{,}132} \cdot L_d^{0{,}755} \cdot I^{0{,}01}$
II Scheme	$B_{sum} = 9{,}395 \cdot g^{0{,}116} \cdot n^{0{,}101} \cdot L_d^{0{,}569} \cdot F_0^{-0{,}019} \cdot I^{0{,}013}$
III Scheme	$B_{sum} = 1{,}272 \cdot g^{0{,}246} \cdot n^{0{,}237} \cdot L_d^{0{,}674} \cdot F_0^{0{,}127} \cdot I^{0{,}202}$

from the regression analysis, the step function for all three alternative schemes is more adequate (Table 2).

The models were tested for practical application by the accuracy, reliability, and adequacy criteria. Accuracy was estimated by the value of the correlation coefficient ($r > 0.7$) and coefficient of determination ($R^2 > 0.5$) for each distribution channel structure (Table 3).

The results of testing the reliability and adequacy of the models by the Fisher criterion indicate their significance in Table 3. Since the calculated values are higher than the tabulated values, the explained variance is significantly higher than the unexplained variance, so the obtained models can be used to form a rational distribution channel structure and to choose the best strategy for servicing cargo owners in regional supply chains.

The degree of influence of such factors as the number of customers and the volume of customer orders on the total present value costs is non-linear; the function of dependence of total delivery costs on specific indicators has an optimum point at which delivery costs reach their minimum value. Within each alternative structure, by varying individual factors, the areas of rational use of each of the proposed variants of the structure of the product distribution channel are determined.

The dependence of total costs on the volume of customer orders (Fig. 2) shows that the rate of change of the total cost function varies with changes in external factors and internal parameters of the delivery system. It is possible to combine the external and internal parameters of the delivery system so that a rational distribution channel structure can be chosen.

Taking into account the conditions proposed when the challenge of intercity delivery of consumer goods in distribution channels is set, with constant values of the delivery distance of 446.5 km (the average distance of cargo delivery), the lowest costs will be when using the II Scheme of cargo delivery using a terminal with an order volume less than or equal to 2.8 tons. If the order volume exceeds 2.8 tons, the direct delivery of goods is more profitable. The dependence on the total cost of cargo delivery as a function of the volume of customer orders is not monotonic. When using any of the proposed distribution channel structures, the cost function forms an optimum point at which the minimum values of total costs for product delivery are achieved.

The results of the analysis of the dependence of total costs on the number of end consumers and the average distance of delivery of goods show that with an increase in the number of consumers over the entire range of changes in the factor from 20 to 200 units, the rational strategy is to use the distribution channel according to the I Scheme of the structure when organizing delivery by the direct variant. The function of

Table 3. Regression models for alternative structures of distribution channel.

Criterion	Structure of the distribution channel		
	I Scheme	II Scheme	III Scheme
Correlation coefficient (r)	0,918	0,854	0,713
Coefficient of determination (R^2)	0,842	0,729	0,509
Fisher's F-test	14,74	14,03	5,39

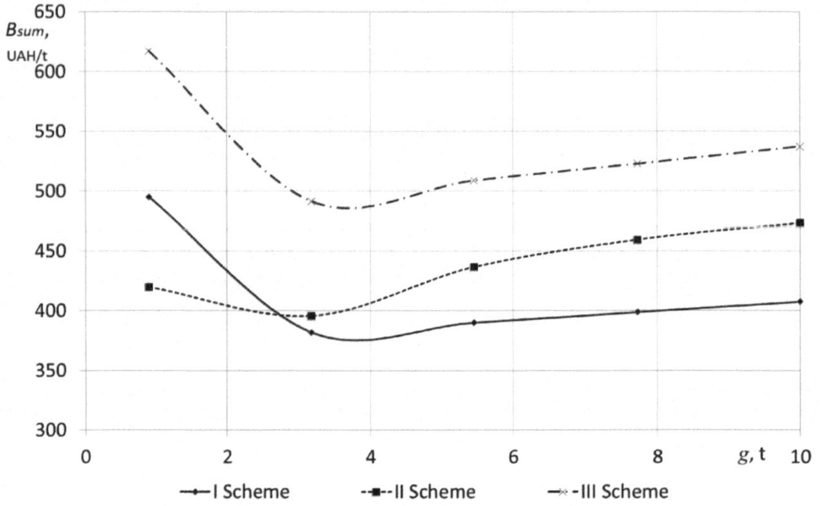

Fig. 2. Dependence of total costs on the volume of cargo ordered by consumers for different distribution channel structures.

total costs, depending on the number of end users, increases evenly. Thus, we can draw the following conclusion: when the factor changes over the entire range, the costs increase evenly by 47%.

The impact of the average delivery distance is monotonous. As the value of this factor increases, the total cost of cargo delivery also increases. The rate of change in total cargo delivery costs as a function of the average distance of cargo delivery is uniform. Thus, it can be concluded that when this indicator increases from 352 to 541 km, the costs increase by 27% evenly. When the delivery distance rises from 352 to 541 km for an order volume of 5.45 tons with several end-user points of 125 units, it is advisable to use the distribution channel according to the first variant (the I Scheme) of the structure over the entire range of parameter changes.

The functions of total costs of product delivery are non-monotonic, as expected in the hypothesis. The rate of change of the total cost function when external factors and internal parameters of the delivery system change are different. By integrating the delivery system's external factors with its internal parameters, one can align the cost functions across various delivery schemes, ensuring uniformity in cost analysis. The intersection point determines the rational areas of use of alternative distribution channel structures, which allows one to choose the strategy for servicing cargo owners in the supply chain.

Based on the analysis, we can offer valuable guidance for choosing a strategy for servicing cargo owners in the transport services market based on the rational use of alternative distribution channels with different structures. In addition, we can identify the leading technical and technological parameters that affect the development of a distribution channel.

Implementing a service strategy based on a rational distribution channel structure increases the efficiency of servicing cargo owners in regional supply chains. Thus, when delivering orders with a volume of more than 2.8 tons under the first scheme, the effect of lowering the total specific cargo delivery costs will equal UAH 25 per ton, or 2.75%.

5 Conclusion

In examining the current state of enhancing the efficiency of services provided to cargo owners, it has been identified that transport enterprises should adopt a holistic approach in choosing their service strategies. This entails not treating material flow, distribution channels, and logistics systems as disparate entities but rather integrating these concepts into a cohesive framework.

Upon analyzing the structure of the research subject, it became clear that the delivery of consumer goods within the context of regional transportation for cargo owners can be effectively managed through various configurations of distribution channels. These configurations should consider the organization of cargo flows and the number of intermediaries involved, ranging from direct delivery options to employing nodes to distribute and consolidate material flows. The parameters that influence the efficiency metric include the minimal total cost for delivering one ton of cargo, the volume of consumer orders, the tally of end-users, and the mean distance over which deliveries are made.

Based on the analysis, we can offer valuable recommendations for choosing a service strategy for cargo owners in the transport services market with conditions similar to the modern Ukrainian market. Delivery operators can use the developed functional models in the study as a decision-support tool for improving the efficiency of servicing cargo owners in regional supply chains under the requirements regarding the quality and cost of services.

The ability to determine the leading technical and technological parameters that influence the choice of a rational distribution channel option allows delivery operators to take into account fluctuations in product demand on the market and reduce delivery costs, which significantly impacts the final price of goods.

Further research is expected to investigate the impact of quality indicators of cargo owner service on the efficiency criterion by supplementing the system of parameters that affect the structure of the distribution channel. This is a relevant area for research in today's changing market conditions and the high degree of customer focus of manufacturers on the needs of their customers.

References

1. Serdaris, P., Antoniadis, I., Tomlekova, N.: Supply chain management: a view of the distribution channel. Bulgarian J. Agr. Sci. **20**, 480–486 (2013)
2. Macioszek, E., Staniek, M., Sierpiński, G.: Analysis of trends in development of freight transport logistics using the example of Silesian Province (Poland) - a case study. Transp. Res. Proc. **27**, 388–395 (2017)
3. Czinkota, M., Kotabe, M., Vrontis, D., Shams, R.: Distribution and supply chain management. In: Springer Texts in Business and Economics Marketing Management, chapter 11, 4th edn, pp. 499–552. Springer, Cham (2021)
4. Naumov, V., Taran, I., Zhanbirov, Z., Mussabayev, B., Konakbai, Z.: Assessing the synergetic effect of selecting the optimal structure of a logistics chain. Sci. J. Silesian Univ. Technol. Ser. Transp. **124**, 109–126 (2024)
5. Muzylyov, D., Shramenko, N., Ivanov, V.: Management decision-making for logistics systems using a fuzzy-neural simulation. In: Cagáňová, D., Horňáková, N., Pusca, A., Cunha, P.F. (eds.) Advances in Industrial Internet of Things, Engineering and Management EAI/Springer Innovations in Communication and Computing 2021, pp. 175–192. Springer, Cham (2021)
6. Pavlenko, O., Muzylyov, D., Ivanov, V., Bartoszuk, M., Jozwik, J.: Management of the grain supply chain during the conflict period: case study Ukraine. Acta Logist. **10**(3), 393–402 (2023)
7. Naumov, V., Shulika, O., Orda, O., Vasiutina, H., Bauer, M., Oliskevych, M.: Shaping the optimal technology for servicing the long-distance deliveries of packaged cargo by road transport. Sustainability. **14**(12), 7283 (2022)
8. Fakhrzad, M., Pourfereidouni, H., Pourfereidouni, M.: Dual-channel supply chain synchronization with deterministic and stochastic demand under cost-sharing contract. Int. J. Supply Oper. Manag. **5**(1), 28–41 (2018)
9. Gündüz, H.I., Hempsch, C.K., Sebastian, HJ. Recent advances in strategic and tactical planning of distribution subnetworks for letter mail. In: Sebastian, HJ, Kaminsky, P., Müller, T. (eds.) Quantitative Approaches in Logistics and Supply Chain Management. Lecture Notes in Logistics, pp. 105–128. Springer, Cham (2015)
10. Nugymanova, G., Nurgaliyeva, M., Zhanbirov, Z., Naumov, V., Taran, I.: Choosing a servicing company's strategy while interacting with freight owners in the road transport market. Nauk. Visn. Nat. Hirn. Univ. **1**, 204–210 (2021)
11. Matsui, K.: Asymmetric product distribution between symmetric manufacturers using dual-channel supply chains. Eur. J. Oper. Res. **248**, 646–657 (2016)
12. Zhang, X., Zhao, Q., Zhang, J., Yue, X.: Logistics service supply chain vertical integration decisions under service efficiency competition. Sustainability. **15**(5), 3915 (2023)
13. Fu, Z., Yue, J., Yan, Y.: Optimized design of multi-level low-carbon logistics distribution scheme based on two stages. Arch. Transp. **69**(1), 145–165 (2024)
14. Pavlenko, O., Muzylyov, D., Trojanowski, P.: Finding a rational option for a cold supply chain using simulation on international routes. In: Arsenyeva, O., Romanova, T., Sukhonos, M., Biletskyi, I., Tsegelnyk, Y. (eds.) Smart Technologies in Urban Engineering. STUE 2023 Lecture Notes in Networks and Systems, vol. 807. Springer, Cham (2023)
15. Jeenanunta, C., Chau, L., Buddhakulsomsiri, J., Dumrongsiri, A.: Determining an optimal warehouse location, capacity, and product allocation in a multi-product, multi-period distribution network: a case study. Int. J. Logist. Syst. Manag. **34**, 510 (2019)
16. Cuda, R., Guastaroba, G., Speranza, M.G.: A survey on two-echelon routing problems. Comput. Oper. Res. **55**, 185–199 (2015)

Optimizing Last-Mile Delivery: Exploring Hybrid Models

Mirna Sofie Jazbec[1] (✉) and Kristijan Rogić[2]

[1] Agent Plus Zagreb d.o.o., Zagreb, Croatia
mirna.jazbec@gmail.com

[2] University of Zagreb, Faculty of Transport and Traffic Sciences, Zagreb, Croatia
kristijan.rogic@fpz.unizg.hr

Abstract. Research on optimizing the last-mile delivery system mostly encompasses a single delivery model. The model is conceived as follows: goods from warehouses located on the outskirts of cities are delivered by commercial delivery vehicles to micro-depots located on the outskirts of city centers or historical cores, from where distribution is carried out either by smaller delivery vehicles or environmentally friendly vehicles, such as small electric vans and/or electric cargo bicycles/tricycles. In the process of selecting the optimal model, various criteria are considered. The aim of the paper is to provide an overview of research in the field of last-mile delivery modeling, with guidelines for future research. The focus is on researching hybrid delivery network models in urban areas, and identifying additional elements that can contribute to greater efficiency in the last-mile delivery system, alongside reviewing existing network optimization criteria. The paper is structured so that the introductory section outlines the purpose of the research. A review of existing research with a critical analysis constitutes the central part of the paper. In conclusion, guidelines for future research in the optimization of hybrid last-mile distribution networks are provided.

Keywords: Optimization · Last-mile delivery · Hybrid model network · Environmentally friendly vehicles · Electric cargo bicycles

1 Introduction

City logistics can be defined as the process for totally optimizing the logistics and transport activities by private companies with support of advanced information systems in urban areas considering the traffic environment, the traffic congestion, the traffic safety and the energy savings within the framework of a market economy [1].

Last-mile delivery is part of city logistics, but it is also a final part of supply chain. Last-mile delivery is most often associated with the delivery of packages to home addresses. But it is also a delivering product to the restaurants, stores and other businesses. The central part of the paper will address challenges regarding last-mile delivery, such as costs, routing, time sensitive deliveries, etc., but also the innovative solutions to achieve greater customer satisfaction. Innovative solutions can be related to

G. Sierpiński et al. (Eds.): TSTP 2024, LNNS 1257, pp. 33–43, 2024.
https://doi.org/10.1007/978-3-031-81777-9_3

vehicles (e.g. electric vehicles), secure deliveries, customer movement (e.g. use of parcel lockers), consolidations strategies, last-lag change (e.g. micro-hubs) [2].

The aim of this paper is to review the existing research and studies regarding optimization of last-mile delivery in urban areas. The paper is structured in four distinct chapters. The first part of the paper is focused on the introduction of the purpose of the research. The central part of the paper contains a review of existing research and studies in different fields of city logistics.

The final part of the paper contains guidelines for future research in the optimization of last-mile delivery, putting the focus on the hybrid model. In this case, hybrid delivery model will combine last-mile delivery using commercial vehicles as well as environmentally friendly vehicles, focusing on electric cargo bikes, in the wider urban areas, not just city center.

2 Research Methodology

Research methodology (Fig. 1) is structured as follows: defining keywords, a comprehensive review of previous research published in the scientific database ScienceDirect of Elsevier, sorting scientific papers according to the title and the year of publication, sorting scientific papers according to relevance, and in the end, reading the papers.

Initial keywords were: 'city logistics'; 'green/electric vehicle'; 'e-commerce'; 'routing'. While reading chosen publications, more keywords appeared, such as "last-mile

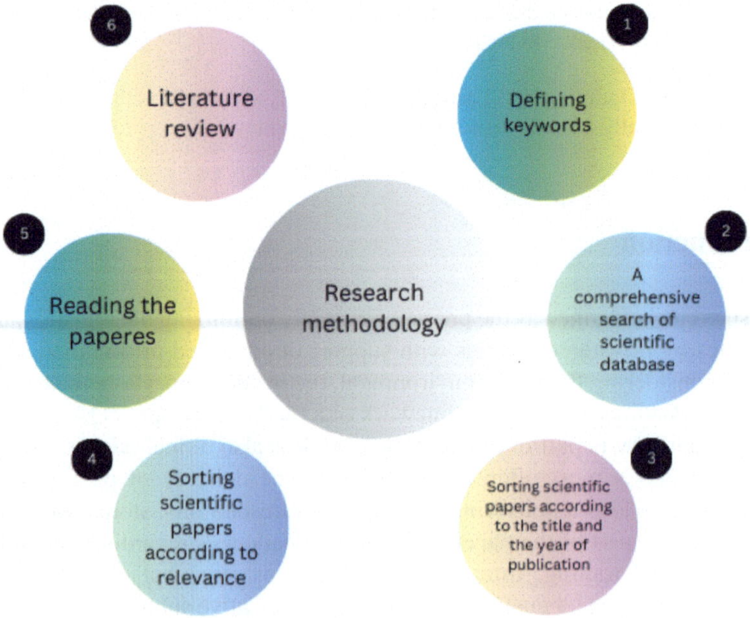

Fig. 1. Research methodology.

delivery"; "innovations"; "hub and spoke"; "electric cargo bicycles" and even "cycle logistics". Most of publications were from recent years, 2014–2024.

3 Last-mile Delivery Solutions and Optimization Models

The first part of the literature review will be focused on the impact of last-mile delivery on the economic, environmental and social aspects and why the need for innovative solutions has arisen. This is followed by the introduction of urban consolidation centers (UCC) and cargo bikes. In the second part, examples from practice will be listed and analyzed, and last-mile delivery routing models.

3.1 Impact of Last-mile Delivery on the Economic, Environmental and Social Aspects

A lot of papers in recent years start with the same fact, the growth of e-commerce. Consequently, there is a growth in the number of deliveries in the cities and the vehicles in the streets [2–4]. There was continuity in the growth, up until the outbreak of virus COVID-19, when e-commerce grew rapidly [2].

The delivery options are no longer driven by the supplier, but more and more influenced by the customers' preferences [18]. Because of the increased demand, to ensure customer satisfaction, service providers had to provide and develop innovative services, including scheduled delivery, redirected delivery, return logistics, same-hour or same day delivery, and such [3]. New services have brought a new set of organizational challenges and in many cases increased operational costs.

Foltyński wrote that cities are growing rapidly. This growth is paired with an increase in residential and business needs. With around 80% of the population living in cities, with the majority in medium-sized cities, this has produced an increase in the movement of goods into the city from external producers [5].

Last-mile delivery is identified as a key bottleneck, e.g. issues related to delivery price, quality of service, tracking/tracing and delivery information, and return logistics [3]. Aljohani et al. states the challenges faced by companies during delivery: narrow streets, lack of parking spaces, undelivered shipments, repeat deliveries, traffic congestion, due to the inability to access the delivery address, a lot of time is wasted carrying packages. Also indicates that service users are demanding, expect same-day or next-day delivery, and prefer direct delivery to e.g. parcel lockers [6].

Also, door delivery has a negative impact on the environment mostly caused by rate of failed deliveries, low vehicle load factor and long travelled distances. Delivery vehicles contribute up to 50% of particulate matter emissions in the cities and are responsible for 40% of transport related CO2 emissions [2]. Afroditi et al. points out environmental concerns and freight management have become vital for companies aiming to reduce energy consumption and greenhouse gas (GHG) emissions. The transport sector is responsible for 30% of the CO2 emissions in the EU while this share reaches 40% in urban areas [7].

There is also social sustainability affected by last-mile deliveries, different logistics companies operating in the same city, generating additional traffic congestion, noise and questioning road safety [2]. According to the European Environment Agency, about 95 million people are exposed to harmful levels of road traffic noise. At least 20% of the urban population is exposed to levels considered harmful to health. In many cities, this percentage can reach 50% of the urban population [19]. In 2020, 40% of all road fatalities in the EU27 occurred on a road inside urban areas according to European Road Safety Observatory (2022) [20].

One of the points of research is sustainable city framework; on the one hand there are public authorities aiming to reduce the transportation impact to increase the attractiveness of their city for inhabitants and visitors (mainly environmental and social sustainability e.g. reduction of congestion, pollution, accidents), while private companies seek to operate at the lowest cost but high-quality and short time in order to satisfy the customers' expectations in a highly competitive market (mainly economic sustainability) [11].

Silva et al. proposed a 4-level hierarchical model to support decision-making in selecting the most adequate last-mile solution within the context of a historic center, considering the triple bottom line of sustainability. The environmental criteria encompass air pollution, noise pollution, and visual pollution; the economic criteria comprise operational costs, load capacity, and delivery speed; and the social criteria comprise congestion, safety & security, and acceptance. Environmentally oriented decision-makers prefer cargo bikes, while decision-makers who prioritize economic and social factors prefer parcel lockers [9].

3.2 Introduction of Urban Consolidation Centers (UCC) and Cargo Bikes

Consolidation strategies to improve last-mile delivery have been investigated since the early 2000s with the concept of urban Consolidation Centers (UCC). UCC are logistics facilities located close to the city center from which consolidated deliveries are made with small vans, avoiding presence of heavy vehicles in city center [2]. Foltyński described the use of UCC as a logistics facility for the last-mile collection and distribution of goods which is situated close to the urban area that it serves. UCC main freight operations are handling of cargo, loading/unloading, warehousing and added value services [5]. Another term, Micro Distribution Centers (MDC) used in literature by Clausen et al. (2016), it has the same characteristics as UCC. The idea of MDCs is to separate the distribution into activities inside and outside the city [8].

According to Gogas and Nathanail the UCCs play a key role with the supply chain, as they constitute the nodal points where the cargo assembled is transshipped from large and polluting heavy good vehicles (HGVs) and freight transport units, to smaller, cleaner – environmental friendly and more flexible ones performing the last-mile distribution in urban areas, in order to protect urban and restricted areas from traffic congestion and impact on environment, safety and quality of life. They proposed UCC concept evaluation including following impact area: economy and energy (energy, development, costs), transport & mobility (transport system, vehicle, infrastructure and technology), society (living standards), policy and measure maturity (awareness level), user uptake (stakeholder approval, knowledge and experience transfer) and environment (air quality, GHG emissions, noise) [17].

Van Heeswijk et al. tested 1458 schemes, measuring both financial and environmental KPIs. Significant environmental improvements were achieved with a UCC, reducing truck kilometers in the city by about 65% and emissions by about 70%. The main challenge was finding financially sustainable schemes. The UCC can achieve the highest revenues by convincing carriers to outsource stops and selling value-adding services to city receivers, the key merits of the UCC are value-adding services such as temporary storage, waste collection, e-tailing logistics, home deliveries, and specialized services such as splitting pallets into smaller loads or putting clothes on hangers. Without monetizing environmental costs, the UCC concept is financially unviable without supporting measures. Temporary subsidies to carriers and a zone-access fee are the most effective measures for achieving profitability and sustainability without external funding. However, financial stability is highly challenging and may be unattainable for some UCCs, depending on user base and operational costs [16].

Schliwa et al. describes the term cycle logistics as the use of human-powered or electrically assisted standard bicycles, cargo bikes and cargo tricycles for the transport of goods between A and B, primarily in urban areas. Cycle logistics includes services of last and first mile logistics and same day deliveries [11].

Foltyński stated number of market barriers for electric vehicles, including electric cargo bikes, related to range, charging time, insufficient charging infrastructure network, high investment costs, lacking or inadequate incentives, lack of information, and disinformation [5]. Disadvantages are limited carrying capacity, and limited distance they can cover, the delivery itself depends on the driver's characteristics, weather, and loading conditions, driver training and making deliveries require a large amount of labor force and investment [10].

Simoni et al. investigated potential implementation of UCCs considering the total costs and the environmental impacts. UCCs can be beneficial as they might reduce transportation in congested areas and increase reliability and efficiency. Moreover, in UCC can be performed some added value services. On the other hand, there are extra costs of developing functional UCC facilities. In the baseline scenario, diesel vans were predominantly used due to their higher capacity and lower fixed costs, resulting in one facility and three van routes. EVs and e-cargo bikes were rarely used due to their limited capacity and service range. Scenario 1 aimed to minimize emissions, leading to the extensive use of e-cargo bikes for last-mile distribution, resulting in higher costs but lower emissions. Scenario 2 introduced the possibility of using UCC with lower costs (local authorities provide public facilities), resulting in three evenly distributed facilities and equal utilization of EVs and e-cargo bikes, leading to lower overall operational costs but did not significantly reduce emissions. Scenario 3 involved subsidizing EVs and e-cargo bikes, which led to reduced emissions while maintaining relatively low distribution costs. Scenario 4 focused on using more environmentally friendly vehicles for the first leg of delivery, leading to significantly reduced emissions while maintaining distribution patterns and costs similar to the baseline scenario. Overall, the study demonstrated trade-offs between costs and emissions in last-mile distribution, highlighting the importance of considering different scenarios and vehicle types to achieve optimal results [12].

The analysis of UCC according to Clausen et al. includes three main aspects: economic aspect evaluates the impact of UCC on the traffic and costs of last-mile delivery, such as a reduction in the number of deliveries or kilometers traveled, environmental aspect emphasizes the saving of emissions, and social aspect assesses the attractiveness of UCC pilot projects for recipients and suppliers [8].

Cargo bikes reduce pollution and congestion, can be driven on a normal road and on bicycle lanes, have access to parts of the city where other delivery vehicles struggle or cannot access at all, reduces delivery costs because there are no parking fees, congestion costs and fuel, it increases the level of security in cities [10].

3.3 Implementation of UCC and Cargo Bikes in Practice

Sochor et al. describes innovative projects in the cities of Stockholm and Barcelona. In the city of Stockholm two solutions were presented, first use of parcel lockers and second micro-hubs for last-mile delivery of refrigerated goods. Project showed it is possible to reduce distance travelled and negative impacts while maintaining time and delivery frequency through adoption of this approach. Although the additional leg of transportation to the micro hub/terminal remains a factor that reduces productivity, it is worth emphasizing that the positive impacts should be seen to outweigh the importance of this issue; the wider socio-economic benefits are likely to be greater than the marginal costs imposed by use of a micro-hub [4].

In Barcelona the use of micro-hubs with last-mile delivery with cargo bikes was presented. They also tested two cargo-bike box innovations, one for regular parcels and one for refrigerated goods. The boxes are detachable from the cargo bikes and are large enough to directly accept goods loaded on a Euro-size pallet. This saves handling time at the hub (by enabling loading a pallet into the box in 1 min instead of 30 min for manually loading individual packages) and means that larger volumes of goods can be transshipped and delivered. With a demonstration of this innovation, parcels could be delivered around 14% quicker, and evidence of increased volumes of goods handled at the Barcelona micro-hubs is expected [4].

Foltyński also describes UCC in Lucca, Italy, where goods are delivered by long–/mid-range freight operators, sorted and delivered to their final destination. Deliveries were made by a fleet of fully electric vehicles. Luccaport was making more than 120 daily deliveries working with 100% load factor and making about 15 deliveries every trip, corresponding to 15% of city deliveries and allowing a reduction in the number of commercial vehicles accessing the historic center of 44%. Luccaport was owned by Lucca Holding, i.e. Lucca city council [5].

Schliwa et al. describes two case studies, Outspoken Delivery and Last-mile Leeds. Approximately 70% of Outspoken's deliveries and collections are pre-scheduled, they collect post from Royal Mail sorting offices, because Royal Mail can't guarantee to deliver it by 9 am. They gained contracts with lager delivery companies, like TNT, which take their items to Outspoken's edge-of-city-center depot for Outspoken to perform last-mile delivery [11].

Last-mile Leeds performs last-mile deliveries on cargo bikes, for the global courier DHL. Due to the city center location of Last-mile Leeds' premises, the maximum delivery distance is approximately a mile and a half, which enables them to offer DHL a very competitive price. Because of this economic advantage, DHL have replaced one of the two vans they operate in Leeds with Last-mile Leeds' services. DHL treats Last-mile Leeds as a subcontractor and pays per piece, to cater flexibly with rising or falling delivery volumes, a common practice in a sector with high seasonal fluctuations [11]. They are still operating in the cities Leeds and Manchester.

Within Paris pilot, 3 MDC were opened and for last-mile delivery were used 2 EVs and 30 electrically assisted cargo tricycles with a range of more than 20 km at a speed of 20 km/h and capacity up to 250 kg and 2.5 m^3. When calculating the optimal route, the following parameters were considered: speed, service time, distance, weight and specific bike accessibility. Data collected during the pilot period showed that the number of round trips increased, but their trip length decreased significantly. This reduction in journey length, combined with the use of cargo bikes, has led to a 30% reduction in total distance traveled and a significant reduction in emissions and noise, and from a social perspective, increased employability and better working conditions [8].

The city of Brussels, faced with an increase in the volume of delivered goods and an increasing number of delivery vehicles, launched a pilot project to encourage the use of alternative delivery methods, including the establishment of UCC. The concept of UCC included the consolidation of goods in a warehouse, near the city center, in the morning and onward delivery to city stores on the same day. Low-emission vehicles such as CNG and small EURO V Diesel vans were used for delivery, while electric scooters and cargo bikes were considered but found to lack sufficient capacity. During the trial phase, which lasted 4 weeks, approximately 1200 deliveries were made with over 2400 goods collected and delivered in 78 round trips. The introduction of UCC led to a 6.3% reduction in deliveries and a significant reduction in total distance traveled, with an overall vehicle occupancy rate of 84.4%. Emissions have also decreased significantly [8].

3.4 Last-Mile Delivery Routing Optimization

Afroditi et al. focuses on modified vehicle routing problems using electric vehicles. Given a homogeneous fleet of depot-returning electric capacitated vehicles and a set of charge stations where vehicles recharge their battery, the goal is to design a set of routes that satisfy the delivery requirements of a set of geographically scattered customers. Each customer has a known demand for delivery, and it must be serviced within a predefined time window. Vehicles are charged at the depot and can travel for a predefined distance before they need to visit a charging station to be fully recharged. Each customer is visited only once by one vehicle and the load of the vehicle must not exceed its maximum capacity. The primary objective is to minimize the fleet size, while the secondary objective is to minimize the distance traveled since the goal is to minimize the total routing cost. The most significant factors affecting the competitiveness of the commercial use of electric vehicles are route feasibility, minimum fleet size, minimum distance traveled, charging level, procurement costs and planning horizon. In addition, the capacity of the vehicle and the time window for delivery to the customer was considered [7].

Caggiani et al. made a model that aims at minimizing the total costs of two echelons considering travel costs, initial vehicles' investment costs, drivers' salary costs, and micro-depot cost. The proposed model highlights the advantages of using e-cargo bikes in restricted traffic zones in terms of energy and investment cost savings. The first echelon is related to the urban zone, while the second echelon is related to the restricted traffic zone and connection is transshipment point (micro-depot) in which e-vans are delivering goods to the e-cargo bikes, according to the request of customers located in restricted traffic zones. According to numerical application results, the proposed solution benefits logistics companies with a high number of daily requests. The use of e-cargo bikes in restricted traffic zones minimizes total costs [13].

The research develops a framework for assessing and ranking zones for cargo cycle operations, focusing on road network conditions (density, linkage ratio, circuitry, and topography) and the coverage of cargo cycle customers. Accessibility, combining land-use and freight demand factors, is measured by the accessibility ratio [10].

A routing optimization model was used to evaluate the reduction in operational cost and GHG emissions when fueled cargo trucks or motorcycles are replaced by LEFV (i.e., e-trikes or e-bikes) in the last-mile distribution in large Latin American cities. Results showed that for the case of a commercial area in Bogota, and two lines of business (i.e., parcel delivery and dry chain), the use of LEFV reduces more than 50% operational cost and more than 95% GHG emissions. These high reductions are due to the synergistic effect of switching from fuel to electricity and reducing the energy consumption of vehicles [14].

The geography of cities plays a major role for the viability of cargo cycles as an alternative to motorized vehicles. High density urban areas, as well as narrow streets in historical city centers, naturally contribute to the attractiveness of cycle logistics. Where this is not the case, urban governance measures regarding traffic regulation orders, e.g. reducing through traffic during the day, can incentivize a modal shift [11]. The topography and slopes of the urban zones are essential features in the CB context. Fewer slopes incentivize CB movements [15].

4 Conclusion and Future Research

Most of the reviewed papers, point out the fact of population growth in urban areas, leading to an increase in deliveries and freight traffic in general, the number of vehicles on the road, congestion, delays, and other problems arising from urban expansion. Because of significant growth in the number of realized shipments, regulation of last-mile delivery became a concern for local authorities and urban residents. Last-mile delivery has a significant negative impact on the environment and quality of life in urban areas, necessitating innovative solutions to minimize this impact as much as possible.

The most common solutions include the introduction of urban consolidation centers on the outskirts of city centers (historic cores), as a central point for deliveries by using electric cargo bikes or small electric vehicles, and/or parcel lockers, where customers handle the last-mile delivery by picking up their packages from designated locations. However, users still prefer home delivery over collecting packages themselves.

According to the reviewed papers, the introduction of consolidation centers has been researched since the early 2000s to reduce the presence of heavy freight vehicles in urban areas. In the examples mentioned in the paper, the implementation of UCCs and delivery with smaller and/or electric vehicles often significantly reduced the negative environmental impact of last-mile delivery. However, the biggest challenge is finding a financially viable model. One of the ideas to make UCCs more financially viable is value-added services. Another challenge is routing electric vehicles considering the development of infrastructure and charging stations. There are many successful examples in practice, in larger cities such as London, Barcelona, and Paris.

A proposal for future research is to develop a hybrid last-mile delivery model and extend the use of electric cargo bikes, or electric driven delivery vehicles to the entire city. A hybrid model involves the use of commercial vehicles in combination with electric vehicles, including cargo bikes. The city could be divided into zones, where not every zone would have its own consolidation center, as investment and maintenance costs would likely be too high. In this case, larger delivery vehicles could serve as consolidation centers for certain zones.

It would be necessary to determine the positions of consolidation centers and the routes where it is most cost-effective to use electric cargo bikes, small delivery vehicles, and/or standard commercial vehicles, considering the characteristics of the vehicles, capacity, infrastructure, charging stations, city topography, and potential weather conditions. City topography is considered because there are cities or parts of cities with steep inclines, and the question is whether electric cargo bikes are cost-effective in such cases. Regarding weather conditions, due to climate changes, the aim is to create a model that quickly adapts to changing conditions. If it rains, there should, for example, be the ability to quickly and efficiently replace electric cargo bikes with smaller electric vans.

The feasibility and profitability of this model could be valued not only through its commercial effects but trough ecologic and spatial effects, in order to reduce the negative impact of the last-mile delivery on the environment and quality of life for the residents, taking in account minimal possible costs for the operators.

References

1. City Logistics: Network Modelling and Intelligent Transport Systems, Eiichi Taniguchi, Russell G. Thompson, Tadashi Yamada, Ron van Duin; Pergamon, Oxford, 2001, ISBN 0 08 043903 9, 260 pp, Journal of Transport Geography **10** (2), 2002, pp 158–159, https://doi.org/10.1016/S0966-6923(01)00041-2
2. Calabrò, G., Le Pira, M., Giuffrida, N., Fazio, M., Inturri, G., Ignaccolo, M.: A spatial agent-based model of e-commerce last-mile logistics towards a delivery-oriented development. Transp. Res. Interdiscip. Persp. **21**, 100895 (2023). https://doi.org/10.1016/j.trip.2023.100895

3. Ducret, R.: Parcel deliveries and urban logistics: changes and challenges in the courier express and parcel sector in Europe — the French case. Res. Transp. Bus. Manag. **11**, 15–22 (2014). https://doi.org/10.1016/j.rtbm.2014.06.009
4. Sochor, J., Fenton, P., de Bode, A., de Azpiazu, M.C., Borraz, R.L., Gasparin, P.: HALLO: hubs for last mile delivery solutions in Barcelona and Stockholm. Transp. Res. Proc. **72**, 3300–3307 (2023). https://doi.org/10.1016/j.trpro.2023.11.797
5. Foltyński, M.: Electric fleets in urban logistics. Procedia. Soc. Behav. Sci. **151**, 48–59 (2014). https://doi.org/10.1016/j.sbspro.2014.10.007
6. Aljohani, K., Thompson, R.G.: Last mile delivery activities in the city centre – Insights into current practices and characteristics of delivery trips. Transp. Res. Proc. **46**, 261–268 (2020). https://doi.org/10.1016/j.trpro.2020.03.189
7. Afroditi, A., Boile, M., Theofanis, S., Sdoukopoulos, E., Margaritis, D.: Electric vehicle routing problem with industry constraints: trends and insights for future research. Transp. Res. Proc. **3**, 452–459 (2014). https://doi.org/10.1016/j.trpro.2014.10.026
8. Clausen, U., Geiger, C., Pöting, M.: Hands-on testing of last mile concepts. Transp. Res. Proc. **14**, 1533–1542 (2016). https://doi.org/10.1016/j.trpro.2016.05.118
9. Silva, V., Amaral, A., Fontes, T.: Towards sustainable last-mile logistics: a decision-making model for complex urban contexts. Sustain. Cities Soc. **96**, 104665 (2023). https://doi.org/10.1016/j.scs.2023.104665
10. Ding, Y., Wang, X.C., Pérez-Guzmán, S., Wojtowicz, J., Conway, A.: Multi-criteria assessment and ranking framework for the potential of cargo cycle operation: using New York city as an example. Transp. Res. A Policy Pract. **179**, 103898 (2024). https://doi.org/10.1016/j.tra.2023.103898
11. Schliwa, G., Armitage, R., Aziz, S., Evans, J., Rhoades, J.: Sustainable city logistics — making cargo cycles viable for urban freight transport. Res. Transp. Bus. Manag. **15**, 50–57 (2015). https://doi.org/10.1016/j.rtbm.2015.02.001
12. Simoni, M.D., Bujanovic, P., Boyles, S.D., Kutanoglu, E.: Urban consolidation solutions for parcel delivery considering location, fleet and route choice. Case Stud. Transp. Policy. **6**(1), 112–124 (2018). https://doi.org/10.1016/j.cstp.2017.11.002
13. Caggiani, L., Colovic, A., Prencipe, L.P., Ottomanelli, M.: A green logistics solution for last-mile deliveries considering e-vans and e-cargo bikes. Transp. Res. Proc. **52**, 75–82 (2021). https://doi.org/10.1016/j.trpro.2021.01.010
14. Díaz-Ramírez, J., Zazueta-Nassif, S., Galarza-Tamez, R., Prato-Sánchez, D., Huertas, J.I.: Characterization of urban distribution networks with light electric freight vehicles. Transp. Res. Part D: Transp. Environ. **119**, 103719 (2023). https://doi.org/10.1016/j.trd.2023.103719
15. Carlos, A., Gonzalez-Calderon, J.J.P.-H., Granada-Muñoz, C.A., Moreno-Palacio, D.P., Arcila-Mena, G.: Cargo bicycles as an alternative to make sustainable last-mile deliveries in Medellin, Colombia. Case Stud. Transp. Policy. **10**(2), 1172–1187 (2022). https://doi.org/10.1016/j.cstp.2022.04.006
16. van Heeswijk, W., Larsen, R., Larsen, A.: An urban consolidation center in the city of Copenhagen: a simulation study. Int. J. Sustain. Transp. **13**(9), 675–691 (2019). https://doi.org/10.1080/15568318.2018.1503380
17. Gogas, M.A., Nathanail, E.: Evaluation of urban consolidation centers: a methodological framework. Proc. Eng. **178**, 461–471 (2017). https://doi.org/10.1016/j.proeng.2017.01.089
18. Perboli, G., Brotcorne, L., Bruni, M.E., Rosano, M.: A new model for last-mile delivery and satellite depots management: the impact of the on-demand economy. Transp. Res. E Logist. Transp. Rev. **145**, 102184 (2021). https://doi.org/10.1016/j.tre.2020.102184

19. European Environment Agency (2022), access URL: https://www.eea.europa.eu/en/topics/in-depth/noise?activeTab=fa515f0c-9ab0-493c-b4cd-58a32dfaae0a
20. Freya Slootmans, European Road Safety Observatory Facts and Figures – Urban areas – 2022, access URL: https://road-safety.transport.ec.europa.eu/system/files/2022-07/ff_roads_inside_urban_areas_20220707.pdf

Assessing the Operating Efficiency and Resilience of Urban Intersections: Case Studies from Poland

Nazanin Zare[1] (✉), Maria Luisa Tumminello[1], Elżbieta Macioszek[2], and Anna Granà[1]

[1] Department of Engineering, University of Palermo, Palermo, Italy
nazanin.zare@unipa.it; marialuisa.tumminello01@unipa.it; anna.grana@unipa.it

[2] Department of Transport Systems, Traffic Engineering and Logistics, Silesian University of Technology, Katowice, Poland
elzbieta.macioszek@polsl.pl

Abstract. This paper presents a heuristic method to evaluate road intersection performance in urban areas, emphasizing resilience during critical events. The study analyzes pre- and post-event control delay variations at three intersections in Poland, categorized as roundabouts, signalized, or two-way stop-controlled intersections. Efficiency and resilience were evaluated by comparing delay times before and after storms. While traffic signals mimicked stop signs post-disaster, no conclusive optimal intersection configuration was identified due to varying location-specific factors such as size, traffic patterns, and network position. The findings highlight the need to integrate resilience into intersection assessments to guide the selection of resilient intersection types, ensuring stable operation amidst events caused by either natural phenomena or human activities. Future research is required on automating performance metric estimations and conducting corridor-level analysis to promote cooperative driving systems.

Keywords: Urban intersections · Efficiency assessment · Resilience analysis · Transportation infrastructure

1 Introduction

Transportation infrastructure is crucial for trade, commerce, and daily activities, making its vulnerability a significant concern for stakeholders. Aligned with Agenda 2030, the Sustainable Development Goals (SDGs), particularly SDG 11, stress the need for resilient cities and communities. This underscores the importance of adaptive, sustainable urban road infrastructure to address environmental and social challenges [1–3]. Resilience in urban systems refers to maintaining functions amid disturbances, adapting to changes, and transforming to enhance capacity [4]. This highlights the importance of quickly returning to equilibrium and facing future challenges. For road infrastructure, resilience means regaining serviceability after disruptions to the

G. Sierpiński et al. (Eds.): TSTP 2024, LNNS 1257, pp. 45–59, 2024.
https://doi.org/10.1007/978-3-031-81777-9_4

transportation system [4–6]. Efficiency in transport infrastructure refers to moving people or goods effectively and reliably while minimizing time, costs, energy, and environmental impacts. Balancing resilience and efficiency in transportation infrastructure upgrades and mitigation activities is crucial but challenging, as the trade-offs between these factors are not yet fully understood [4, 7–9].

Existing literature encompasses numerous studies evaluating resilience to disasters like floods, storms, and earthquakes, employing diverse methodologies such as indicators, indices, and conceptual frameworks. The study in [2] scrutinizes recurrent congestion as an internal disruption on urban roads, analyzing spatial-temporal traffic patterns compared to other freeways and signal-controlled arterial roads. Adaptive traffic signal control strategies enhance resilience during traffic disruptions by adjusting control plans to reduce overall delay times [2, 4, 10–13]. Despite these advancements, few studies have concentrated on the urban road network nodes, predominately highlighting resilience at city or network levels [2, 8]. Traditional road designs, prioritizing efficiency, may falter during unforeseen events. Emphasizing resilience integration into transportation planning and road infrastructure design, while comprehending resilience-efficiency trade-offs, presents persistent challenges.

The study enriches the scientific comprehension of intersection project assessment and optimization, unveiling the interaction among intersection design, control strategies, and urban resilience. By incorporating resilience into efficiency evaluations of urban intersections, it fills a gap in the literature, thereby enhancing considerations for transportation infrastructure design and management.

1.1 Aim of the Study

Acknowledging the significance of efficiency and resilience in urban intersections, this paper seeks to integrate resilience into the evaluation of their operational efficiency. The key questions addressed include: "How can urban intersections be structured and regulated to improve efficiency and resilience, particularly in response to storm-related disruptions?", and "Do different types of traffic control affect the efficiency and resilience of isolated intersections, given identical traffic volumes, location, or geometry?"

The study underscores the necessity of a flexible assessment method that incorporates both resilience and efficiency. Building upon prior study [2, 6], the proposed approach provides a versatile method applicable to various intersection designs, aiding in their assessment and optimization. It involves evaluating delay times at roundabouts, signalized, and two-way stop-controlled (TWSC) intersections pre- and post-storms, revealing significant alterations in traffic signal delay times across the analyzed case studies. These results offer valuable guidance for future traffic management and planning strategies, aiding in selecting intersection types that maintain stability during disruptions and enhance resilience. Furthermore, the study provides insights into the relationship between intersection design, control strategies, and urban resilience, thereby enhancing considerations for transportation infrastructure planning and management.

Following the introduction and research aim, Sect. 1.2 presents a systematic literature review to elucidate the study's purposes and identify research gaps. Section 2 introduces the study area, case studies, data collection, and research methodology. Subsequently, Sect. 3 presents and discusses the analysis results. The final section contains the study's conclusions and potential future developments.

1.2 Literature Review

A literature review focused on urban road intersections from a resilience perspective, explored assessment methods, highlighting key factors, synergies, and challenges in transitioning environments. Inspired by the step-based method in [2, 14], our search focused on studies that blended efficiency and resilience in assessing the operating efficiency of urban intersections for a total of 36 papers, as shown in Fig. 1. These research studies are clustered into three themes, as shown in Table 1, and briefly outlined in the following sections.

The themes in Table 1 are presented briefly in the following parts:

Road Design and Resilience durable materials, proactive design features like alignment, cross-section, and urban structure enhance road resilience, improving recovery and urban safety [15, 17–20, 22, 26]. Integrating intelligent transportation systems and human-AI (Artificial intelligence) collaboration is crucial for mitigating disruptions and emphasizing urban sustainability and efficiency [16, 23–25, 27, 29]. Advanced traffic signals, medians, and stop signs at intersections can comprehensively reduce the risks of pedestrian fatality [28]. Research is essential to pinpoint design features enhancing road resilience and enabling swift recovery from disruptions, given inadequate assessment tools [15, 17, 18, 22, 26, 27]. Optimizing urban intersections for resilience involves focusing on dynamic interactions and spatio-temporal characteristics [21, 26, 27]. Integrating resilience into transportation systems enhances sustainability and mitigates disruptions in cities [23, 24, 28, 29]. Comprehensive methodologies are

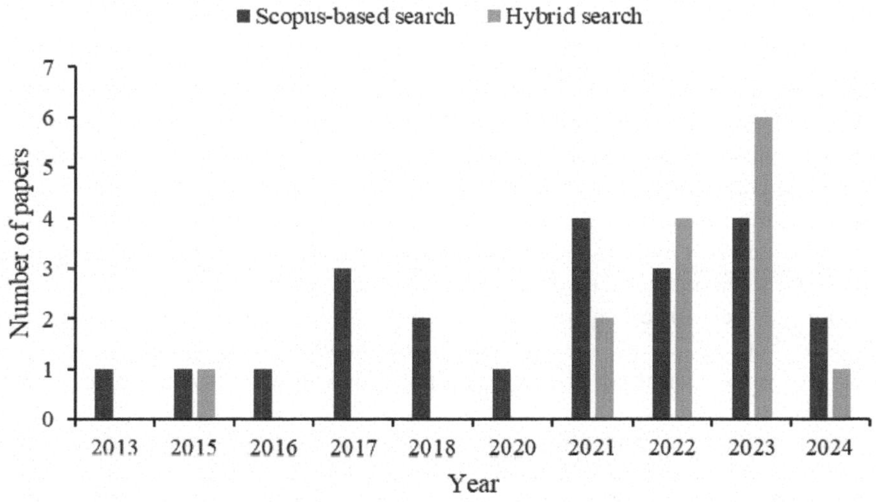

Fig. 1. The number of papers from Scopus and the hybrid search method.

Table 1. Identified key points from the literature.

Theme	Search[a]	No. of studies	References
Road design and resilience	Scopus	7	[15–21]
	Hybrid	8	[22–29]
Traffic management and smart systems	Scopus	11	[11, 30–42]
	Hybrid	5	[43–47]
Studies on urban mobility and society	Scopus	3	[48–50]
	Hybrid	2	[51, 52]

[a]Peer-reviewed papers were identified in the Scopus database through searches of titles, abstracts, and keywords relevant to the topic. Furthermore, the Hybrid search enabled the inclusion of additional references by manual searches of journal websites.

necessary to integrate resilience into road and intersection performance assessments [16, 19, 24, 25, 28].

Traffic Management and Smart Systems which, in turn, can be divided into four sub-themes: 1) Autonomous Driving and Connected and Autonomous Vehicles (CAV) s, including highlights advancements and challenges in autonomous driving, proposing frameworks for legal, ethical, and behavioral considerations [32]. Additionally, it discusses CAVs' impact on roundabouts [38, 42], resilient navigation [41], mixed traffic control [45], urban traffic control [35], and secure CAV communication [39]; 2) Physical Infrastructure, concerning the AVs' impact on physical infrastructure that includes geometric design changes, pavement deterioration, bridge safety concerns, parking lot issues, and the need for new facilities [25, 36]; 3) Urban Transport and Mobility, focusing on the improvement of urban transport with vehicle telematics for safety, environmental, and sustainability issues [31]. Additionally, it covers topics such as traffic speed prediction in smart cities [34], energy efficiency for electric CAVs [44], and big data for road surface analysis [30]; 4) Resilience and Safety that introduces metrics and frameworks for urban resilience and safety. It includes discussions on congestion quantification [11], spatio-temporal intersection evaluation [21], resilience-based stormwater management [22], infrastructure criticality during disasters [40], and urban-safety resilience with GIS [47].

Studies on Urban Mobility and Society The potential of connected and autonomous vehicles to enhance road safety is significant, yet there are socio-economic risks such as increased congestion and urban sprawl [50]. The concept of the '15-minute city' is promoted to improve urban resilience and well-being, advocating for disaster resilience frameworks and soft mobility benefits [43, 51]. Furthermore, community engagement is crucial for sustainable and resilient urban environments, with studies highlighting the pandemic's impact on urban mobility [48, 49, 52]. Emphasizing the importance of addressing both social and environmental sustainability, these studies advocate for public involvement in urban planning and policymaking to enhance urban resilience [48, 49, 52].

Current methods of evaluating resilience in urban road infrastructure face barriers such as limited incorporation of local community knowledge [11, 14], constraints in providing comprehensive resilience metrics [10, 23, 28], and challenges in design considerations for elevation, road grading, and drainage systems to cope with heavy rainfall or flooding [10]. There is a need for a better understanding of how urban communities conceptualize their past and future resilience [10, 15]. Enhancing resilient road infrastructure involves focusing on the physical layout and geometry of roads and intersections to improve endurance and recovery from disruptions like storms [13, 52]. Connected infrastructure and Intelligent Transportation Systems can optimize traffic management and control signals at intersections [15, 53] but their implementation is challenged by high costs [17, 23], connectivity issues in remote areas [39], and the need for collaborative efforts from technology developers, policymakers, regulatory bodies, and the public.

However, specific metrics are needed to measure resilience pre and post disruptions, alongside understanding how intersection design affects efficiency and resilience [47]. Future research should consider diverse intersection types supporting vulnerable road users [26], and how people envision their future concerning resilience assessment. The development of a heuristic approach to evaluate the efficiency of road intersections from a resilience standpoint represents a significant contribution to filling these gaps.

2 Materials and Research Methods

The case studies and data collection along with the methodology used to assess the efficiency and resilience of three urban road intersections' layouts in Poland are presented in this section.

The case studies and the data collection are explained in Sect. 2.1, while the performed methodology is described in Sect. 2.2.

2.1 Overview of Case Studies and Data Collection

According to the aims of this paper, three intersections operating in urban road networks of Poland in two different cities were identified. Two of these intersections are signalized intersections located in Katowice, Poland, the other one is a roundabout located in Siemianowice Śląskie, Poland. These intersections are assumed to be isolated along with the aim of this research. Figure 2 illustrates the views and geometric sketches of these case studies.

This paper extends our previous research [2, 6] and examines three case studies involving four-leg intersections situated in Poland, encompassing both signalized and roundabout intersections. Case studies 1 and 2 are signalized intersections in an urban area, while intersection 3 is a roundabout. Data were collected by videotaping the traffic situation during peak hours from September to November 2023, along with field surveys. The geometries of the current and assumed configurations of case studies are illustrated in Table 2. All the intersections comply with Polish regulations [54–56].

Traffic flow graphs of the case studies are presented in Fig. 3. The total traffic flows for intersections 1, 2, and 3 are 1452, 1099, and 1311 vehicles per hour, respectively.

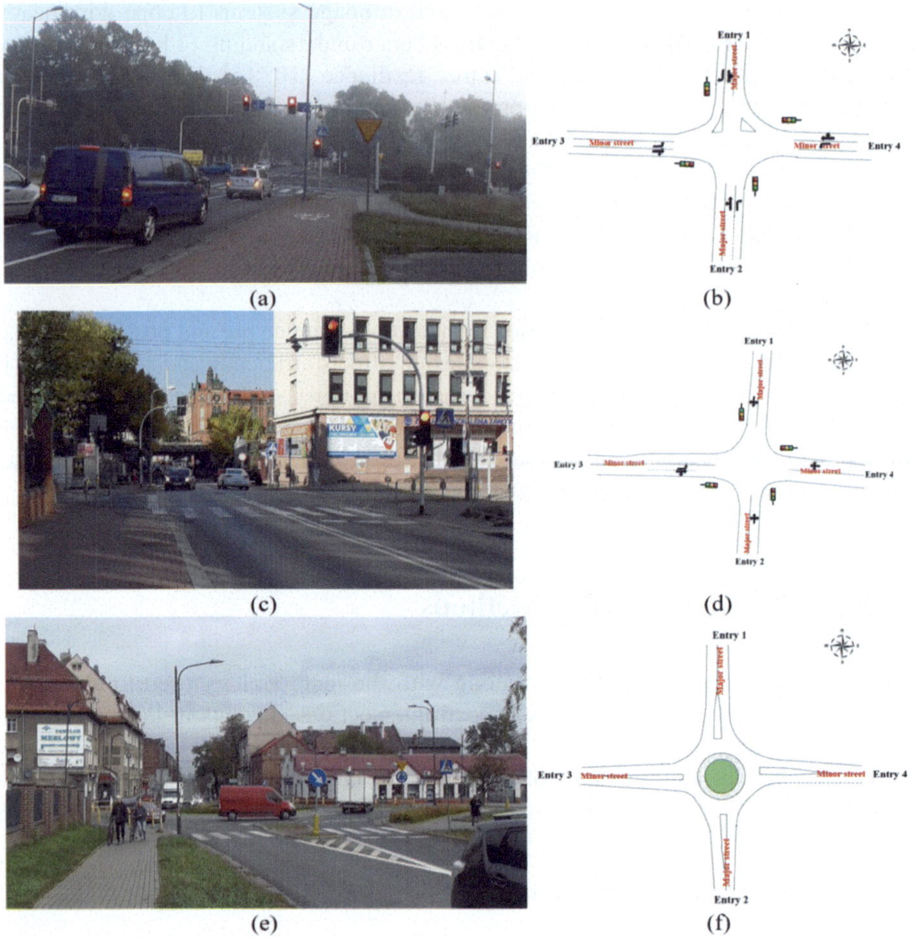

Fig. 2. The selected intersections in Poland: (a) A view from the south entry at intersection 1 (latitude: 50.251557, longitude: 19.029674); (b) The geometric sketch of intersection 1; (c) A view from the southeast entry at intersection 2 (latitude: 50.255549, longitude: 19.030781); (d) The geometric sketch of intersection 2; (e) A view from the south entry at intersection 3 (latitude: 50.319716, longitude: 19.003237); (f) The geometric sketch of intersection 3.

Heavy vehicles constituted 3%, 2%, and 12% of the total volume in case studies 1, 2, and 3, respectively.

Although case study 3 is a roundabout that can support U-turn movements, observations of U-turn movements during field surveys were very limited, particularly evident during peak hour flows at the intersection (see Fig. 3c).

Case studies 1 and 2 are signalized intersections with a cycle length of 80 s each. For the hypothetical traffic light in case study 3, the cycle length is assumed to be 90 s.

In case study 1, the cycle length is divided into four phases. In phase 1, a green time of 14 s is given to all movements on entry 1. Similarly, in phase 2, a green time of 14 s is given to all movements on entry 2. In phase 3, a green time of 20 s is given to the

Table 2. The geometries of the current and assumed configurations of case studies.

Case study	No. of entry (exit) lanes	Road type[a]	Entry (Exit) lane width (m)	Width (No.) of roundabout circulatory roadway (m)
1	2 (1)	Major street	3.5 (3.5)	3.5 (1)
	2(1)	Minor street	3.5 (3.5)	
2	1(1)	Major street	3.5 (3.5)	3.5 (1)
	1(1)[b]	Minor street	3.5 (3.5)[c]	
3	1(1)	Major street	3.5 (3.5)	5.0 (1)[d]
	1(1)	Minor street	3.5 (3.5)	

[a]The major and minor streets shown in Fig. 2(a), (c), and (e), are differentiated based on the number of entry lanes, the amount of entry (exit) traffic or control mode.
[b]One-lane entries and exits except for the entry from the west direction street which has two entry lanes.
[c]3.5 meters width for entries and exits except for the entry from the west direction which has 2.5 meters width in each lane.
[d]The central island of the roundabout has a traversable apron.

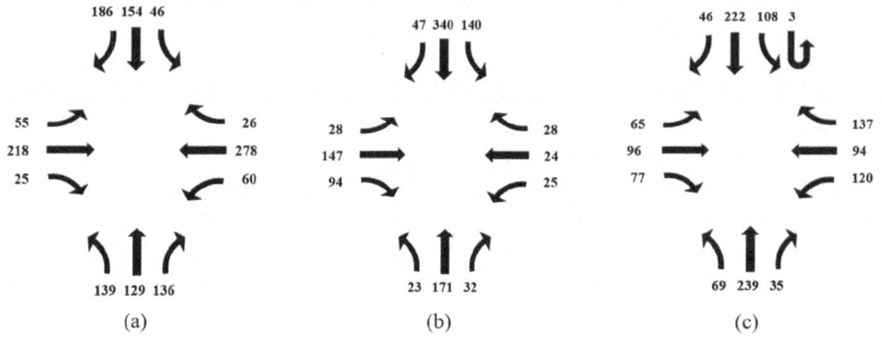

Fig. 3. Traffic flow graphs (veh/h): (a) case study 1, (b) case study 2, & (c) case study 3.

right-turn and through movements on entries 3 and 4. In phase 4, a green time of 23 s is given to the left-turn movements of entries 3 and 4, and the right-turn movements of entries 1 and 2.

The cycle length in case study 2 is divided into two phases. In phase 1, a green time of 53 s is given to the major street approaches (entries 1 and 2). In phase 2, a green time of 20 s is given to the minor street approaches (entries 3 and 4).

In case study 3, the cycle length is assumed to be divided into two phases. In phase 1, a green time of 40 s is given to the major street approaches (entries 1 and 2). In phase 2, a green time of 40 s is given to the minor street approaches (entries 3 and 4).

2.2 Methodology

In alignment with the paper's objectives, this analysis considered two distinct scenarios: pre-storm, representing normal conditions before any storms, and post-storm, where the storm has passed but its effects, such as power outages, persist.

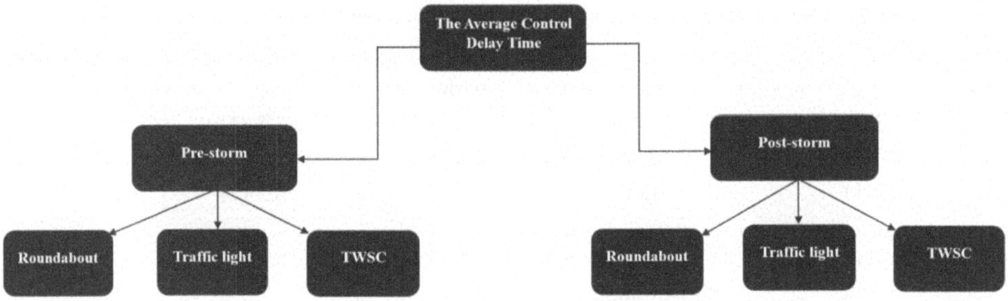

Fig. 4. The studied situations and traffic control modes.

Figure 2(a), (c), and (e) provided the baseline for our analysis, assuming each intersection functions as either a roundabout, a signalized, or a TWSC intersection before and after the storm.

Figure 4 summarizes the situations studied and the traffic control modes in the conceptualized scenarios. Intersection efficiency and resilience were evaluated by analyzing the average control delay times of roundabouts, traffic signals, and TWSCs before and after critical events. This assessment assumed that traffic signals function as stop signs post-disaster due to electrical failure. The study investigated three traffic control modes across two scenarios.

The hypothesis of the applied methodology in this study is to consider the average control delay time to compare the efficiency and resilience of each intersection. The methodology for the average control delay times calculation of intersections draws from [9]. The control delay reflects the time vehicles spend slowed or stopped at intersections due to traffic control devices, and it is a key metric for intersection level of service (LOS) [9]. Here, the total control delay times of the examined intersections are used to compare different traffic control types. The equations for calculating the control delay time can be found in [6].

We compared the average delay times of roundabouts, signalized, and two-way stop-controlled intersections to assess their relative efficiency in typical circumstances. In a post-storm scenario, roads may be blocked, and traffic lights could operate as stop signs. Signalized intersections were assumed to operate similarly to two-way stop-controlled intersections following disasters. Additionally, the average delay times of roundabouts, signalized, and two-way stop-controlled intersections were compared to assess their resilience in disaster situations. The change in post-storm delay times was quantified to comprehend intersection efficiency during disasters, thus indicating their resilience.

The total control delay times are usually used only when comparing the control delay between different traffic control types [9]. Thus, we analyze the total control delay time of the examined case studies to evaluate the efficiency and resilience of roundabout, signalized, and two-way stop-controlled intersections before and after a disaster event.

3 Results and Discussion

In this section, the results are presented. The average control delay time is considered as the index to assess the performance efficiency and resilience trade-offs of the examined case studies.

The comparison between control delay time at an intersection for three types of traffic control modes is studied in this paper. Figures 5, 6 and 7 illustrate these results. Table 3 presents the average delay time and the level of service for each approach in the case studies, both before and after a storm.

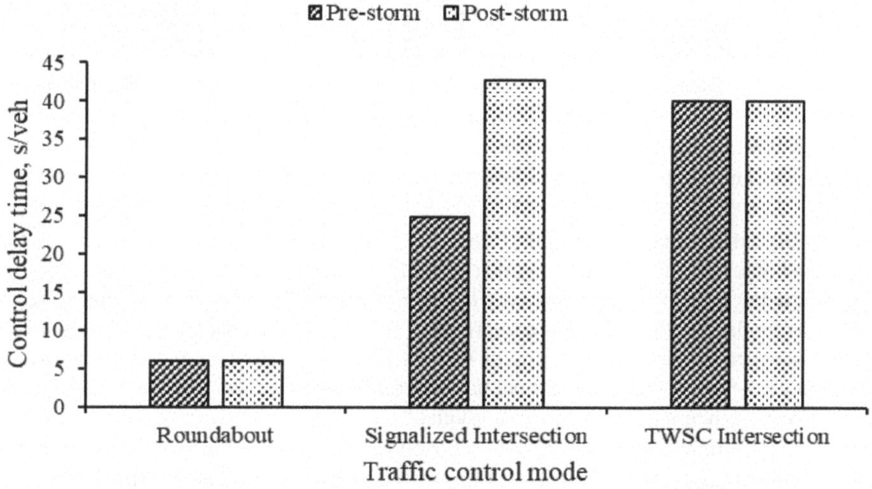

Fig. 5. The comparison of performance efficiency and resilience in case study 1.

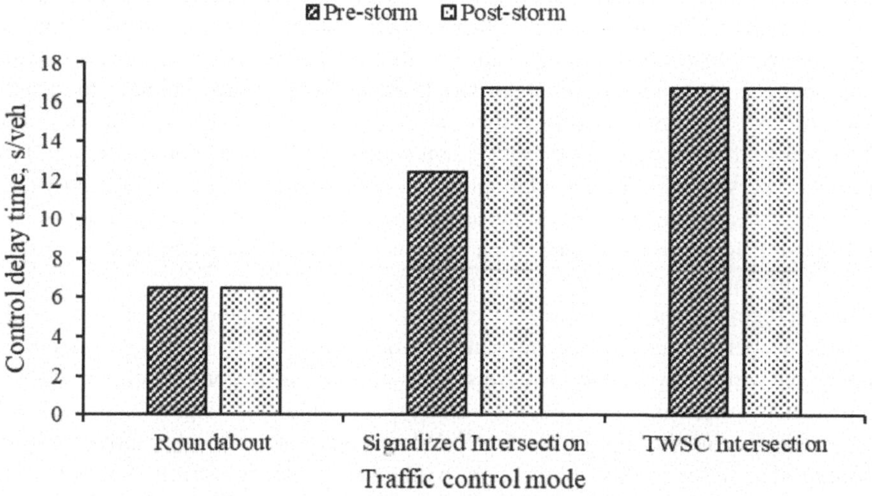

Fig. 6. The comparison of performance efficiency and resilience in case study 2.

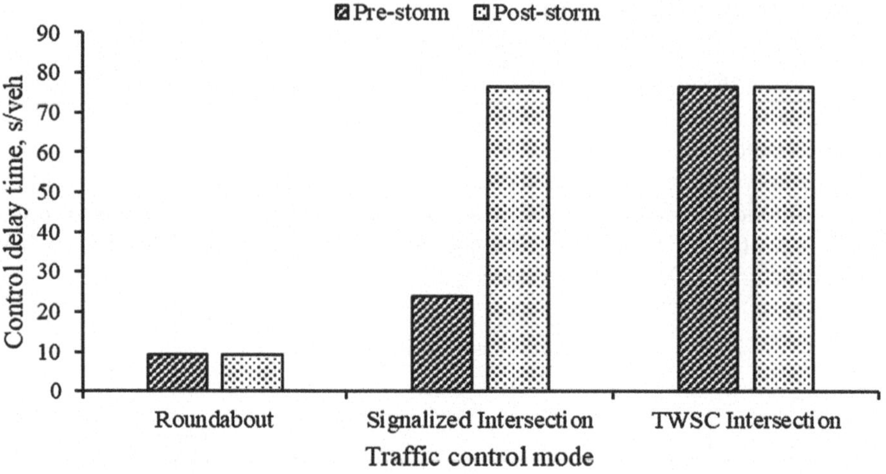

Fig. 7. The comparison of performance efficiency and resilience in case study 3.

The results indicate that in case study 1, the current traffic control mode, which is traffic light, yields a level of service of "C", considered acceptable in a dense road network based on the average control delay time of the intersection. However, if the traffic of this intersection were controlled by a roundabout, the LOS would improve to "A". In contrast, analyzing the intersection as a TWSC intersection may result in a LOS of "E". Efficiency rankings, based on the average control delay time, show that a roundabout has better performance than a signalized intersection, with the TWSC counterpart being the least efficient.

Similarly, resilience assessments, also based on the average control delay time, confirm that the roundabout is the most resilient alternative, followed by the TWSC, and the signalized intersection being the least resilient (see Fig. 5).

To assess the performance efficiency and resilience of the intersection in case study 2 (see Fig. 2(c)), the average delay times of the intersection in normal and critical situations are studied. This intersection's current traffic control type is the traffic lights, and the results show that it operates efficiently with the LOS of "B". Alternatively, assuming the intersection to be a roundabout would result in an improvement in the level of service to "A". However, defining this intersection to be controlled by stop signs (TWSC), would result in the LOS of "C", which can also be considered efficient. From a resilience perspective, the roundabout is the most resilient alternative, followed by TWSC and the signalized intersections as the second most resilient traffic control types (see Fig. 6). It is worth mentioning that these results were predictable due to the field surveys confirming this LOS for the current situation.

The average control delay time of case study 3, a roundabout, is also studied to evaluate the performance efficiency and resilience in pre- and post-storm circumstances. The results show that this case study's current traffic control setup is the most efficient and resilient among all other studied types (roundabout, signalized, and TWSC intersections), achieving the LOS of "A". When considering the alternative of implementing traffic lights in this intersection, while it is an efficient traffic control type, it would lack resilience, as indicated by a comparison of the average control delay times

Table 3. The average control delay time and LOS per approach.

Case study	Situation	Type of Traffic control	Control delay (s/veh) and LOS			
			Entry 1[a]	Entry 2[a]	Entry 3[a]	Entry 4[a]
1	Pre-storm	Roundabout	6.35(A)	5.60 (A)	5.48 (A)	6.55 (A)
		Traffic light	22.64 (C)	26.98 (D)	23.34 (C)	26.27 (D)
		TWSC	0.19 (A)	0.37 (A)	88.49 (F)	86.66 (F)
	Post-storm	Roundabout	6.35(A)	5.60 (A)	5.48 (A)	6.55 (A)
		Traffic light	4.51 (A)	6.24 (A)	88.49 (F)	86.66 (F)
		TWSC	0.19 (A)	0.37 (A)	88.49 (F)	86.66 (F)
2	Pre-storm	Roundabout	7.35 (A)	5.98 (A)	7.22 (A)	4.12 (A)
		Traffic light	10.26 (B)	6.12 (A)	25.72 (D)	25.04 (D)
		TWSC	8.25 (A)	7.51 (A)	38.92 (E)	18.50 (C)
	Post-storm	Roundabout	7.35 (A)	5.98 (A)	7.22 (A)	4.12 (A)
		Traffic light	8.25 (A)	7.51 (A)	38.92 (E)	18.50 (C)
		TWSC	8.25 (A)	7.51 (A)	38.92 (E)	18.50 (C)
3	Pre-storm	Roundabout	9.37 (A)	8.08 (A)	9.04 (A)	9.81 (A)
		Traffic light	27.08 (D)	23.54 (C)	19.50 (C)	23.97 (C)
		TWSC	2.99 (A)	2.04 (A)	194.46 (F)	148.18 (F)
	Post-storm	Roundabout	9.37 (A)	8.08 (A)	9.04 (A)	9.81 (A)
		Traffic light	2.99 (A)	2.04 (A)	194.46 (F)	148.18 (F)
		TWSC	2.99 (A)	2.04 (A)	194.46 (F)	148.18 (F)

[a]Entries 1–4 are introduced in Fig. 2(b), (d), (f).

(see Fig. 7). On the other hand, a TWSC scenario would lack efficiency and resilience in case study 3 due to the intersection's average control delay time.

These results demonstrate that in pre-storm situations, the control delay times per approach for roundabouts and signalized intersections remain consistent across the entire intersection. However, due to the inherent characteristics of TWSC intersections (e.g. priority rules), the control delay times vary significantly between major and minor street approaches. In post-storm situations, the control delay times per approach for roundabouts remain consistent across the entire intersection, whereas, for signalized and TWSC intersections, they vary considerably between major and minor street approaches.

4 Conclusions

Building upon previous research [6], this paper employs a new heuristic method to evaluate road intersection performance in urban areas, emphasizing resilience. Inspired by [8], the study explores pre- and post-event control delay time variations to enhance the understanding of intersection efficiency during critical events. The analysis begins by examining normal (pre-storm) and post-storm scenarios, using three urban intersections in Poland as case studies. Considering intersection size and traffic patterns, each was presumed to operate as either a roundabout, a signalized, or a two-way stop-controlled intersection before and after the storm.

Intersection efficiency was evaluated by comparing roundabouts, traffic signals, and stop signs before storms, while resilience was assessed by comparing these configurations after storms. As traffic signals imitate stop signs following disasters, changes in delay times were observed in the analyzed case studies before and after storms. However, no definitive conclusion could be drawn regarding the optimal intersection configuration or control mode. Factors specific to each site, such as size, traffic patterns, and position within the road network, must be carefully considered before selecting the appropriate type. Therefore, the paper presents a summary of findings, indicating areas for further investigation into intersection design and operation to identify the most resilient alternative among the examined types.

Additionally, it stresses integrating resilience into intersection assessments over their lifespan. As future roads require greater adaptation and resilience, this research guides the selection of intersection types to ensure stable operation amidst natural phenomena or human activities. Societally, it explores facilitating information sharing among road users and operators to coordinate actions for implementing cooperative driving systems.

Recognizing the constraints of the present findings to particular intersection types and underlying assumptions, future research avenues could encompass automating the estimation of intersection performance metrics using turning movement data and examining operations at a corridor level. This aims for a broader system analysis that accounts for geometric delays in smart intersections.

Acknowledgements. This publication is supported by the Rector's Pro-Quality Grant, Silesian University of Technology, Poland, grant number 12/040/RGJ24/0064, and Silesian University of Technology grant number BK-285/RT4/2024, 12/040/BK_24/0065. The authors acknowledge the support of the Sustainable Mobility Center (Centro Nazionale 1133 per la Mobilità Sostenibile-CNMS) under Grant CN00000023 CUP B73C22000760001.

References

1. Assarkhaniki, Z., Sabri, S., Rajabifard, A., Kahalimoghadam, M.: Advancing sustainable development goals: embedding resilience assessment. Sustain. Sci. **18**(5), 2405–2421 (2023)
2. Zare, N., Macioszek, E., Granà, A., Giuffrè, T.: Blending efficiency and resilience in the performance assessment of urban intersections: a novel heuristic informed by literature review. Sustainability. **16**, 1–24., paper ID 2450 (2024)
3. Sustainable Development Goals. United Nations. SDG Indicators, Metadata repository, https://unstats.un.org/sdgs/metadata/
4. Therrien, M.C., Normandin, J.M., Paterson, S., Pelling, M.: Mapping and weaving for urban resilience implementation: a tale of two cities. Cities. **108**, 1–9 (2021)
5. Espinet, X., Schweikert, A., van den Heever, N., Chinowsky, P.: Planning resilient roads for the future environment and climate change: quantifying the vulnerability of the primary transport infrastructure system in Mexico. Transp. Policy. **50**, 78–86 (2016)

6. Zare, N., Macioszek, E., Granà, A.: A heuristic approach to assess the performance efficiency of road intersections in urban environments from the resilience perspective. In: Macioszek, E., Jurdana, I., Sierpiński, G. (eds.) Road Traffic Analysis, Theoretical Approaches and Practical Solutions. LNNS, vol. 877, pp. 17–26. Springer, Cham (2024)
7. Tumminello, M.L., Macioszek, E., Granà, A., Giuffrè, T.: A microsimulation-based modelling approach for connected and automated vehicles on roundabouts. In: Macioszek, E., Granà, A., Sierpiński, G. (eds.) Advanced Solutions and Practical Applications in Road Traffic Engineering. LNNS, vol. 604, pp. 49–68. Springer, Cham (2023). https://doi.org/10.1007/978-3-031-22359-4_4
8. Bengigi, E.: Efficiency and Resilience Trade-Offs for Roadway Intersection Design in the US Virgin Islands. Naval Postgraduate School, Monterey (2020)
9. National Academies of Sciences, Engineering and Medicine: Highway Capacity Manual, Six Edition: a Guide for Multimodal Mobility Analysis. Transportation Research Board, Washington (2016)
10. Andronov, R., Leverents, E.: Calculation of vehicle delay at signal-controlled intersections with adaptive traffic control algorithm. MATEC Web Conf. **143**, 1–6 (2018)
11. Tang, J., Heinimann, H.R.: A resilience-oriented approach for quantitatively assessing recurrent spatial-temporal congestion on urban roads. PLoS One. **13**(1), 1–22 (2018)
12. Tong, P.: Characteristics, dimensions and methods of current assessment for urban resilience to climate-related disasters: a systematic review of the literature. Int. J. Disaster Risk Reduct. **60**, 1–10 (2021)
13. Yao, K., Chen, S.: Percolation-based resilience modeling and active intervention of disrupted urban traffic network during a snowstorm. J. Transp. Eng. A Syst. **149**(5), 1–15 (2023)
14. Berrang-Ford, L., Pearce, T., Ford, J.D.: Systematic review approaches for climate change adaptation research. Reg. Environ. Chang. **15**, 755–769 (2015)
15. Dia, H.: Urban mobility: challenges and opportunities. In: Low Carbon Mobility for Future Cities: Prinicples and Applications. The Institution of Engineering and Technology, London (2017)
16. Ilbeigi, M., Meimand, M.E.: Empirical analysis of impacts of post-disaster human mobility patterns on the resilience of transportation networks. In: Construction Research Congress 2020 Proceedings. American Society of Civil Engineers Reston (2020)
17. Juremalani, J., Chauhan, K.A.: Evaluation of use of non-motorized vehicles for shopping trips under mix traffic conditions. In: ASCE India Conference 2017. Urbanization Challenges in Emerging Economies: Energy an Water Infrastructure, Transportation Infrastructure, and Planning and Financing Proceedings. American Society of Civil Engineers (2017)
18. Kirova, N.: Pedestrian Flow: Monitoring and Prediction. Regulation (2016), https://papers.cumincad.org/data/works/att/acadia20_84.pdf
19. McDaniels, T.L., Chang, S.E., Hawkins, D., Chew, G., Longstaff, H.: Towards disaster-resilient cities: an approach for setting priorities in infrastructure mitigation efforts. Environ. Syst. Decis. **35**, 252–263 (2015)
20. Nikitas, A., Tsigdinos, S., Karolemeas, C., Kourmpa, E., Bakogiannis, E.: Cycling in the era of Covid-19: lessons learnt and best practice policy recommendations for a more bike-centric future. Sustainability. **13**(9), 1–25 (2021)
21. Zhao, Z., Tang, L., Yang, X., Zhang, H., Li, G., Li, Q.: Identifying critical urban intersections from a fine-grained spatio-temporal perspective. Travel Behav. Soc. **34**, 1–18 (2024)
22. Hettiarachchi, S., Wasko, C., Sharma, A.: Rethinking urban storm water management through resilience-the case for using green infrastructure in our warming world. Cities. **128**, 1–12 (2022)

23. Mageto, J., Twinomurinzi, H., Luke, R., Mhlongo, S., Bwalya, K., Bvuma, S.: Building resilience into smart mobility for urban cities: an emerging economy perspective. Int. J. Prod. Res., 1–18 (2022) https://www.tandfonline.com/doi/epdf/10.1080/00207543.2022.2139866?needAccess=true
24. Mattinzioli, T., Butt, A.A., Harvey, J.: Literature review on pavements and electric vehicle interaction: a research roadmap. Transp. Res. D Transp. Environ. **122**, 1–10 (2023)
25. Othman, K.: Impact of autonomous vehicles on the physical infrastructure: changes and challenges. Designs. **5**(3), 1–30 (2021)
26. Pratelli, A., Leandri, P., Aiello, R., Souleyrette, R.R.: Intersection redesign for network resilience and safety. Promet-Traffic Transp. **33**(2), 297–308 (2021)
27. Shafiei-Dastjerdi, M., Lak, A.: Towards resilient place emphasizing urban form: an assessment framework in urban design. Sustain. Cities Soc. **96**, 1–17 (2023)
28. Tanishita, M., Sekiguchi, Y., Sunaga, D.: Impact analysis of road infrastructure and traffic control on severity of pedestrian-vehicle crashes at intersections and non-intersections using bias-reduced logistic regression. IATSS Res. **47**(2), 233–239 (2023)
29. Yang, P.P.J.: Energy resilient urban form: a design perspective. Energy Procedia. **75**, 2922–2927 (2015)
30. Bianchini, D., De Antonellis, V., Garda, M.: Relevance-based big data exploration for smart road maintenance. In: Sellami, M., Ceravolo, P., Reijers, H.A., Gaaloul, W., Panetto, H. (eds.) Cooperative Information Systems. LNCS, vol. 13591, pp. 19–36. Springer, Cham (2022)
31. Ghaffarpasand, O., Burke, M., Osei, L.K., Ursell, H., Chapman, S., Pope, F.D.: Vehicle telematics for safer, cleaner and more sustainable urban transport: a review. Sustainability. **14**(24), 1–20 (2022)
32. Li, J., Yu, C., Shen, Z., Su, Z., Ma, W.: A survey on urban traffic control under mixed traffic environment with connected automated vehicles. Transp. Res. C Emerg. Technol. **154**, 1–21 (2023)
33. Tumminello, M.L., Macioszek, E., Granà, A., Giuffrè, T.: Simulation-based analysis of "what-if" scenarios with connected and automated vehicles navigating roundabouts. Sensors. **22**, 1–30 (2022)
34. Magalhaes, R.P., Lettich, F., Macedo, J.A., Nardini, F.M., Perego, R., Renso, C., Trani, R.: Speed prediction in large and dynamic traffic sensor networks. Inf. Syst. **98**, 1–98 (2021)
35. Martínez-Díaz, M., Soriguera, F.: Autonomous vehicles: theoretical and practical challenges. Transp. Res. Poc. **33**, 275–282 (2018)
36. Rondón Quintana, H.A., Reyes Lizcano, F.A., Urazán Bonells, C.F.: Effect of reducing traffic speed on the durability of a road asphalt layer. Ingeniare. **21**(1), 139–149 (2013)
37. Tumminello, M.L., Macioszek, E., Granà, A., Giuffrè, T.: Evaluating traffic-calming-based urban road design solutions featuring cooperative driving technologies in energy efficiency transition for smart cities. Energies. **16**, 1–28., Paper ID 7325 (2023)
38. Shawky, M., Alsobky, A., Al Sobky, A., Hassan, A.: Traffic safety assessment for roundabout intersections using drone photography and conflict technique. Ain Shams Eng. J. **14**(6), 1–10 (2023)
39. Stavdas, A., Kosmatos, E., Maple, C., Hugues-Salas, E., Epiphaniou, G., Fowler, D.S., Razak, S.A., Matrakidis, C., Yuan, H., Lord, A.: Quantum key distribution for V2I communications with software defined networking. IET Quant. Commun. **5**(1), 38–45 (2024)
40. Tariverdi, M., Nunez-Del-Prado, M., Leonova, N., Rentschler, J.: Measuring accessibility to public services and infrastructure criticality for disasters risk management. Sci. Rep. **13**(1), 1–16., paper ID 1569 (2023)
41. Thakker, R., Alatur, N., Fan, D.D., Tordesillas, J., Paton, M., Otsu, K., Toupet, O., Agha-Mohammadi, A.A.: Autonomous off-road navigation over extreme terrains with perceptually-challenging conditions, 1–12 (2021), https://arxiv.org/pdf/2101.11110

42. Tumminello, M.L., Macioszek, E., Granà, A., Giuffrè, T.: A methodological framework to assess road infrastructure safety and performance efficiency in the transition toward cooperative driving. Sustainability. **15**, 1–20. paper ID 9345 (2023). https://doi.org/10.3390/su15129345
43. Abdelfattah, L., Deponte, D., Fossa, G.: The 15-minute city: interpreting the model to bring out urban resiliencies. Transp. Res. Proc. **60**, 330–337 (2022)
44. Faghihian, H., Sargolzaei, A.: Energy efficiency of connected autonomous vehicles: a review. Electronics. **12**(19), 1–22., paper ID 4086 (2023)
45. Li, D., Zhu, F., Wu, J., Wong, Y.D., Chen, T.: Managing mixed traffic at signalized intersections: an adaptive signal control and CAV coordination system based on deep reinforcement learning. Expert Syst. Appl. **238**, 1–18., paper ID 121959 (2024)
46. Macioszek, E., Granà, A., Fernandes, P., Coelho, M.C.: New perspectives and challenges in traffic and transportation engineering supporting energy saving in smart cities-a multidisciplinary approach to a global problem. Energies. **15**, 1–8., paper ID 4191 (2022)
47. Xiang, Y., Chen, Y., Su, Y., Chen, Z., Meng, J.: Research on the evaluation and spatial–temporal evolution of safe and resilient cities based on catastrophe theory-a case study of ten regions in Western China. Sustainability. **15**(12), 1–50., paper ID 9698 (2023)
48. Arimah, B.: Infrastructure as a catalyst for the prosperity of African cities. Proc. Eng. **198**, 245–266 (2017)
49. Tarasi, D., Daras, T., Tournaki, S., Tsoutsos, T.: Transportation in the Mediterranean during the Covid-19 pandemic era. Glob. Trans. **3**, 55–71 (2021)
50. Docherty, I.: The transition to automated mobility: how well do connected and autonomous vehicles really fit into a sustainable transport future? In: Curtis, C. (ed.) Handbook of Sustainable Transport, pp. 209–218. Edward Elgar Publishing, Cheltenham (2020) https://china.elgaronline.com/display/edcoll/9781789900460/9781789900460.00033.xml
51. Cheek, W., Chmutina, K.: Measuring resilience in the assumed city. Int. J. Disaster Risk Sci. **13**(3), 317–329 (2022)
52. Chiroli, D.M.D.G., Menezes, M.G., Zola, F.C., Aragão, F.V., de Almeida, R.D., Tebcherani, S.M.: Integrating resilience and sustainability: a systematic analysis of resilient cities using ISO 37123. Int. J. Disaster Risk Reduct. **96**, 1–16., paper ID 103960 (2023)
53. Tumminello, M.L., Macioszek, E., Granà, A.: Insights into simulated smart mobility on roundabouts: achievements, lessons learned, and steps ahead. Sustainability. **16**, 1–32., paper ID 4079 (2024)
54. Bąk, R., Chodur, J.; Gaca, S.; Kieć, M.; Ostrowski, K.: Guidelines for the design of road intersections. Part 2: Ordinary and channelized intersections (2022), https://www.gov.pl/web/infrastruktura/wr-d
55. General Directorate for National Roads and Motorways: Road Intersection Design Guidelines. Part II Roundabout. General Directorate for National Roads and Motorways, Warsaw (2001)
56. Bąk, R., Chodur, J.; Gaca, S.; Kieć, M.; Ostrowski, K.: Guidelines for the design of road intersections. Part 3 Roundabouts (2022), https://www.gov.pl/web/infrastruktura/wr-d

Optimal Parking Location in Ramadi City Using Analytic Hierarchy Process and GIS Applications

Khalid Hardan Mhana[1], Hamid Ahmed Awad[1], Firas Alrawi[2,3](✉), and Maher Zuhair Mohammed[1]

[1] College of Engineering, University of Anbar, Ramadi, Iraq
Khald.hardan@uoanbar.edu.iq; hamid.awad@uoanbar.edu.iq; Maher.Zuhair@uoanbar.edu.iq

[2] School of Geographical Sciences and Urban Planning, Arizona State University, Tempe, AZ, USA

[3] Urban and Regional Planning Center, University of Baghdad, Baghdad, Iraq
fhamodi@asu.edu; dr.firas@uobaghdad.edu.iq

Abstract. The Analytic Hierarchy Process (AHP), Fuzzy Analytic Hierarchy Process (FAHP), and GIS applications are utilized to assess sixty-three Ramadi districts concerning the ideal parking location. The results of the AHP and FAHP approaches indicate that factors like area, amenities, land price, automobile ownership, traffic volume, and road density have a weighted impact of 21%, 17%, 16%, 27%, 11%, and 8% respectively on the acceptability of parking areas in the 63 districts of Ramadi city. This conclusion can be inferred based on the results obtained from the AHP and FAHP methodologies. The AHP and FAHP approach has identified that 8.07% and 13.31% of the total area is highly suitable for parking. On the contrary, 12.06% and 5.15% of total area is low suitability for parking.

Keywords: Analytic hierarchy process · Fuzzy analytic hierarchy process · GIS · Parking location

1 Introduction

There has been and will continue to be a progressive increase in the global population. As the population continues to grow, there is a corresponding rise in demand. This growth indicates the expansion of cities and towns. The contemporary world is undergoing rapid transformation due to the increase in the population and the migration of people from rural areas to large urban centers [1]. The progression of technology and the sophistication of technology contribute to the growth of urbanization. Increasing the number of people living in metropolitan areas has primarily had a negative influence, manifesting in various forms. As a consequence of this, people are becoming more aware of concerns about the environment and urban problems. As a result of the rapid increase of the population and the development of cities and towns, various challenges

G. Sierpiński et al. (Eds.): TSTP 2024, LNNS 1257, pp. 61–75, 2024.
https://doi.org/10.1007/978-3-031-81777-9_5

arise, which invariably lead to a greater scarcity of resources and services. This, in turn, causes substantial shortages in a variety of disciplines and systems [2]. As a result of this growth, the transportation industry is rapidly expanding, leading to an increase in challenges related to urban transportation [3]. The expansion in the population also affects the growth of the requirement for higher usage of automobiles [4], and as a result, traffic is presented as one of the most significant difficulties that urban transportation faces [5]. As a result of the growing number of vehicles and the associated lack of parking spots, cities are experiencing many negative consequences. In several cities, one of the most significant and significant traffic concerns and problems is the availability of public parking. At the same time that it is one of the most vital urban utilities and one of the most difficult responsibilities, public parking is one of the most significant difficulties that rapidly expanding cities are currently confronting [6]. The issue of parking is a source of frustration for motorists and has developed into a widespread issue in every city [7]. An organized surface or location that is sought for the purpose of parking automobiles is referred to as a parking lot. The process of locating parking places is a significant societal issue that is both stressful and time-consuming [8]. The problem with parking causes traffic congestion, and environmental pollution, causes a variety of concerns to become more severe, and has an effect on road congestion [9]. Parking places vary according to location and characteristics. It is necessary to provide a parking place in order to satisfy the demand for parking. In accordance with the findings of Litman [10] and Demir et al. [9], parking requirements are influenced by a wide range of factors, including but not limited to physical location, land use, demographics, and the availability of public transportation networks. The transportation system is comprised of a multitude of components that have an impact on parking lots. In order to achieve urban land uses and transportation systems, it is extremely important to have easy access to parking facilities and to map parking lots. This requires the employment of complex methodologies and is an extremely important component [6]. In the context of contemporary urban mobility, parking spaces fulfill an important function. The planning of parking continues to be one of the most important aspects, and it is also one of the primary goals that urban planners have in mind when they consider their work to improve the social amenity of the urban atmosphere [11]. Site selection of parking lots is one of the essential decisions that must be made in public urban spaces. This decision needs careful consideration of a wide range of criteria, parameters, and circumstances, and it is typically of a somewhat complicated nature [12]. With the proper selection of new locations for public parking, the efficiency and efficacy of the parking system will undoubtedly be improved, and at the same time, it will be able to absorb a great deal of the traffic problems that have been occurring [13]. Regardless of whether it is an ordinary or parametric invasion, the location must be accessible [14]. Regulations for parking should be allocated by local governments, and they should not be discriminatory. Parking design should assure safety, continuity, and lanes for individuals with disabilities [15]. Nevertheless, this is not sufficient to tackle the challenges that offer distinctive solutions [16], particularly in the process of selecting parking spaces for public use. Parking spots that are efficient lower the amount of traffic and the amount of parking that is marginal. As a Multi-Criteria Decision Analysis (MCDA) problem, the triage of a parking space is also supposed to be solved. In the beginning, such a process begins with identifying and defining the worry, and it culminates in a concluding verdict with a number of different options [17]. In spite of the fact that Geographical Information Systems (GIS) offer a wide range of capabilities in terms of spatial analysis and visualization [18], such a range is not sufficient for the resolution of complicated issues. Therefore, in order to tackle complex tasks of this kind, it is imperative that a method that combines Geographic Information System (GIS) and Analytic Hierarchy Process (AHP) be utilized [19]. At the same time that the integration of information systems is required, processes such as globalization, urbanization, and climate change, in addition to other technological and environmental improvements, require a large amount of attention and thought [20]. The Analytic Hierarchy Process (AHP) technique, which was first presented by Saaty in 1980 [21], has shown to be an effective and widely used method and measurement that is utilized in a wide variety of site selection concerns for the purpose of calculating the weights of criteria [22]. It has been demonstrated that the AHP approach and GIS technologies are useful in resolving site selection issues, and they are also utilized in the process of picking parking places that are suitable [3]. The purpose of this methodology was to locate ideal areas for parking facilities inside the city of Konya Turkey. This study evaluated various factors, such as the area of the district, amenities, land price, car ownership, traffic volume, and road density using the AHP (Analytic Hierarchy Process) and Fuzzy Analytic Hierarchy Process (FAHP) method and Geographic Information Systems (GIS) applications.

2 Method

2.1 Study Area

Ramadi is located in the western part of Iraq, namely at the point where the longitudes (43°–20′) and (43°–12′) east meet, as well as the latitudes (33°–27′) and (33°–23′) north meet. A significant portion of the city may be found in the southern region. The Al-Anbar Governorate is geographically divided by the Euphrates River, with the eastern portion of the governorate being located on one side of the river. The Anbar Statistics Directorate reports that Ramadi encompasses a vast expanse that spans 318,088 square kilometers and has a built area that is 55.934 km^2. Because it is both central and administratively significant, the city serves as a lively hub for activity among the neighborhoods, and it also contains a considerable concentration of major government entities. On top of that, it is a key transportation hub that acts as a point of intersection for several transportation routes, including major highways and international roads like 10 and 12. It is for this reason that many consider it to be an essential access point for passage between the eastern and western parts of Iraq. The strategic placement of the city allowed for uninterrupted transportation throughout the entire year. This provides the city with a regional relevance, which immediately makes it easier for it to advance and expand (Fig. 1).

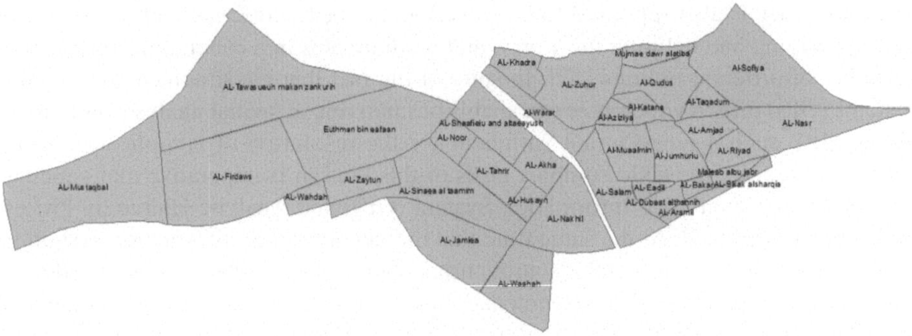

Fig. 1. Ramadi City (Case Study).

Table 1. Selected Criteria.

Criteria	Code AHP	Code FAHP	Definition
District Area	C1	CF1	Total area in each district (km^2)
Amenities	C2	CF2	Number service building in each district
Car Ownership	C3	CF3	Percentage of population that have at least one vehicle in each district
Land Price	C4	CF4	Average price of one square meter in each district (in million)
Traffic Volume	C5	CF5	Average traffic volume in each district (veh/hr)
Road Density	C6	CF6	The area of road network in each district (km^2)

2.2 Criteria

A total of six criteria were utilized in this study to successfully determine suitable locations for parking areas. It is important to take into consideration the current study to determine the factors that have an impact on the parking area. Area, amenities, land price, car ownership, volume of traffic, and road density. These criteria were classified based on the vector and raster datasets and collected from different methods. This section of the document contains in-depth explanations of each criterion as well as the sub-criteria that correspond to it. Furthermore, the maps that exemplify the spatial suitability distribution of each criterion were developed. These maps were constructed. Table 1 displays the outcomes of the sub-criteria that were carried out.

- Area of district

The area of the district might be taken into consideration when deciding where these locations of parking should be constructed. When it comes to classifying the thirty-six districts area a classification system that is known as Natural Breaks (Jenks) is utilized. Where there is a high district the location of the parking area will be more appropriate.

- Amenities

The proximity of the facility is an important factor to consider when selecting parking places. This is especially true when parking lots are located near facilities like shopping centers, hospitals, schools, and government buildings. The reason for this is that parking spots that are located near facilities will be more advantageous than those that are further away. The use of Google Maps is required to generate a map layer that satisfies the amenities criteria. The transformation of the coordinate system is the next step that is carried out after the coordinates of the amenities have been determined. It is more likely that the area will be good for parking if the district has a high concentration of amenities.

- Land price

Since the study area comprises places that have a wide range of land values, land values are taken into consideration as an economic issue when in the process of determining an ideal location for a parking lot. When the land values map is seen from this perspective, the layer that is formed is the one that is stated in the Iraqi Dinar unit. Following the completion of the computation of the average land values for each district, the data are then connected with the geographical criterion layer that is relevant to the scenario. It is more likely that the space will be suitable for the parking area if the land value is low.

- Car ownership

Car ownership can be evaluated in the location selection of the parking area. The map layer is created by using car ownership statistics of the thirty-six districts. This new map layer is classified into five ranges. If the location has a high car ownership, it will be more suitable for parking areas.

- Traffic volume

Another criterion that is taken into consideration while determining the location of parking areas is the average traffic volume in each district. The official company in Ramadi, Iraq, is the source of the spatial and attribute information regarding the traffic volumes used. These data are utilized in the composition of the traffic volume map layer. A location that is more suited for parking areas will have a high traffic volume in the district.

- Road density

To select appropriate sites for parking areas, it is essential to take into consideration the road density that is represented by the volume of traffic. A vector data format is implemented in the creation of the road map layer. A district that has a high road density will be more suited for parking areas because of the increased number of roads in the district.

2.3 Analytical Hierarchy Process (AHP)

The analytical hierarchy process (AHP) is a semi-quantitative method in which selections are maintained using weights through a pair-wise comparison procedure. This method eliminates discrepancies in the decision-making process. The AHP encompasses the subsequent five phases, which include the following: (1) the disruption of choice complications into constituent matters; (2) the groundwork of these criteria in hierarchical teaching; (3) the development of numerical values to regulate the relative location of each criterion version to their meaning; (4) the comparison matrix system; and (5) the normalization of all criteria weights. Numerous investigators also utilized the Analytic Hierarchy Process (AHP) and the Fuzzy-AHP in order to give weight criteria for composed causal criterion and inconsistent causal criterion examples. The establishment comparison matrix, the combination of multiple rulings, and the degree of fuzzy weights consistency are all obscured by the fuzzy-AHP structure process [21].

2.4 Fuzzy Analytical Hierarchy Process (FAHP)

The FAHP approach has been developed with the purpose of surpassing the shortcomings of the classic AHP method and providing new operational answers to problems that arise in unpredictable environments. Similarly, when the number of criteria and substitutes is considerable, the implementation of AHP can produce negative features in terms of the pairwise comparison matrix. Given that it makes use of linguistic variable stars in a pairwise comparison matrix, the FAHP technique was not only appropriate but also more accurate than other methods. In FAHP, a modification was proposed to implement a method that would apply fuzzy numbers with three sides to the pairwise comparison matrix. In order to build a pairwise comparison matrix, the question was occupied by the thought process of the decision makers [22].

2.5 Pairwise Analysis

This matrix compares and contrasts two different criteria that are used in solitary dating. It differentiates the column constituent from the noise component, and it emphasizes the significance of the column constituent in comparison to the approximation row component of the output findings. This was an amount that was achieved today, which is a location strength, and it ranged from approximately 1–9. Take, for example, the assumption that there are two distinct standards, such as X and Y. Uncertainty, the standard X is just as crucial as the standard Y, the location standards may be close to 1, and as a result, the diagonal rudiments of the matrix would always be 1. Standard X is, without a doubt, of far more significance than standard Y, and directory standards might be 9 [21].

2.6 Criteria Weight

At this step, significant consideration is given to separate weighting criteria. In the first place, a standardized eigenvalue matrix is formed. Each matrix standard is then disconnected by the sum of its pillar, and the weight of each row mean in this legalized eigenvalue matrix equals the weight of that particular standard.

2.7 Consistency Ratio (CR)

The AHP can be predicted by using CR, which operates in a relaxed manner through the equation. This assists in classifying the adapted balanced pairwise comparison matrix illogicality that was reported as having come from skilled or knowledge ruling. Any pairwise matrix symbol that has a CR value less than 0.1 is reliable. In accordance with the criterion number that is being utilized, the RI values are a predetermined quantity that is subject to vary.

3 Results

The AHP and FAHP approach was utilized via the process of determining the weights of the criterion in this study. A questionnaire was developed and distributed to professionals working in the subject, including urban planners, engineers, and specialists from both the commercial and public sectors, including municipal employees, in order to collect feedback from those who are already considered experts in the industry. A pairwise comparison of criteria is used in this research to determine the relative importance of each of the criteria on a scale ranging from 1 to 9. This research makes use of the AHP approach. The weights of the criteria are derived based on the geometric mean of the opinions of experienced professionals, including engineers and urban planners, and generates pairwise comparison matrices. The weights of the criteria that were acquired are provided in Table 2.

As can be seen in Table 2, the criterion that carries the most weight is the car ownership (C4), which has a weight of 0.27. On the other hand, the road density requirement (C6) has the lowest weight, which is 0.08. To determine suitable locations for parking area, a GIS-based AHP and FAHP approach is used. All spatial analysis processes for the selected criteria (C1–C6) for both methods (AHP and FAHP) can be seen in Figs. 2, 3, 4, 5, 6, 7, 8, 9, 10, 11, 12 and 13.

Table 2. Criteria Weight.

Criteria	Code AHP	Code FAHP	Weight (%)
District Area	C1	CF1	21
Amenities	C2	CF2	17
Land Price	C3	CF3	16
Car Ownership	C4	CF4	27
Traffic Volume	C5	CF5	11
Road Density	C6	CF6	8

Fig. 2. Area criteria (C1).

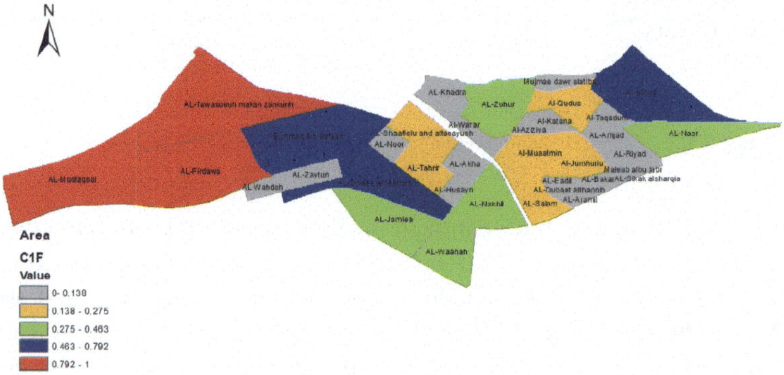

Fig. 3. Area criteria (CF1).

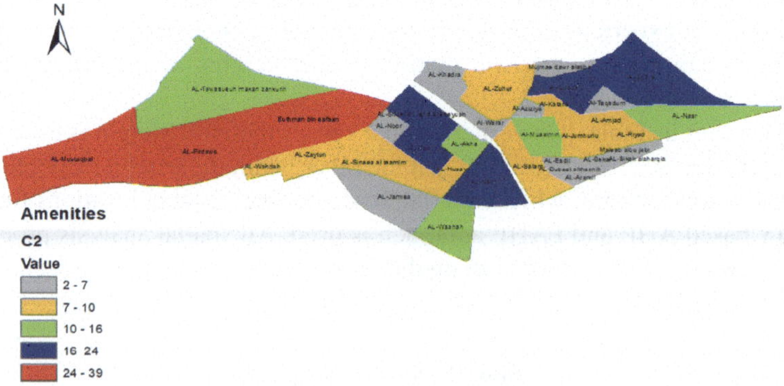

Fig. 4. Amenities criteria (C2).

A GIS-based AHP and FAHP approach is utilized in order to ascertain the most appropriate locations for parking areas in Ramadi districts. Figures 2, 3, 4, 5, 6, 7, 8, 9, 10, 11, 12 and 13 display all of the spatial analysis processes that were performed for the parameters that were chosen (C1 through C6) for both the AHP and FAHP methods.

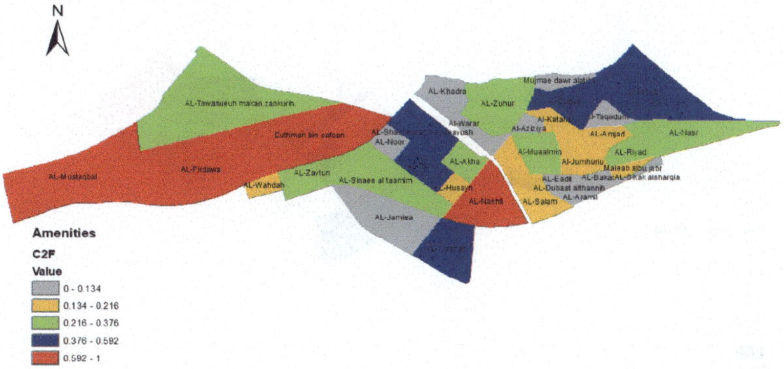

Fig. 5. Amenities criteria (CF2).

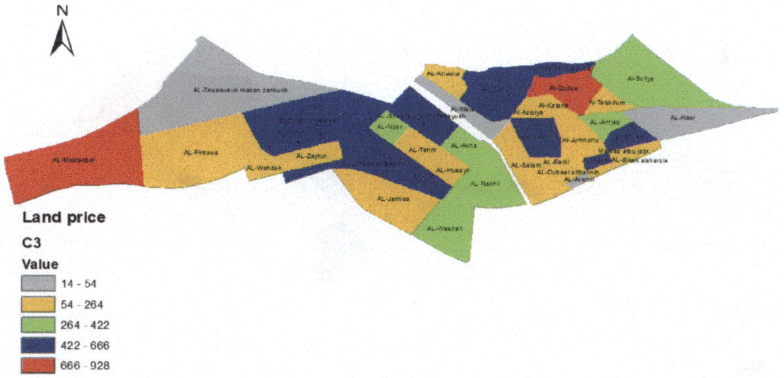

Fig. 6. Land price criteria (C3).

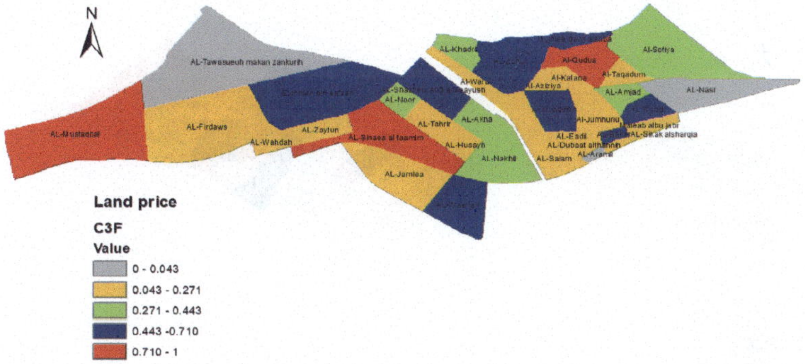

Fig. 7. Land price criteria (CF3).

For all criteria (C1–C6) the spatial analysis maps based on the value of the criteria has been classified into five classifications as follows:

- Gray color represented very low suitability district for parking area.
- Orange color represented low suitability district for parking area.

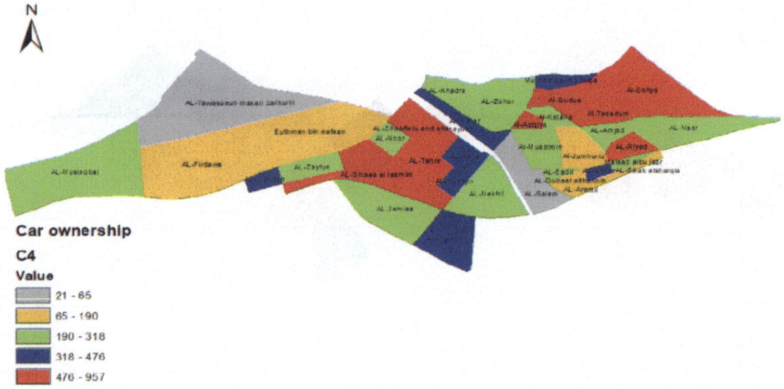

Fig. 8. Car ownership criteria (C4).

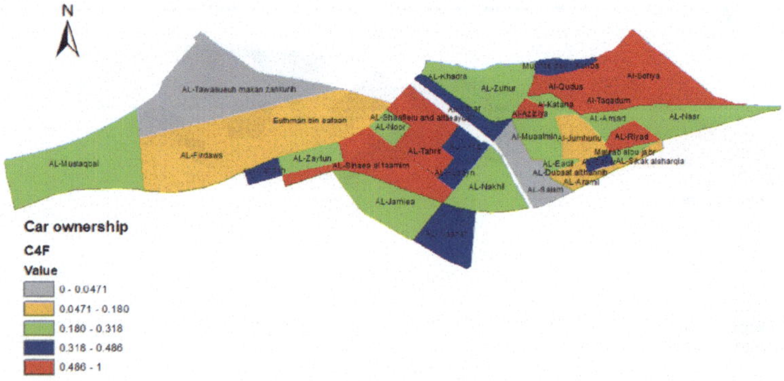

Fig. 9. Car ownership criteria (CF4).

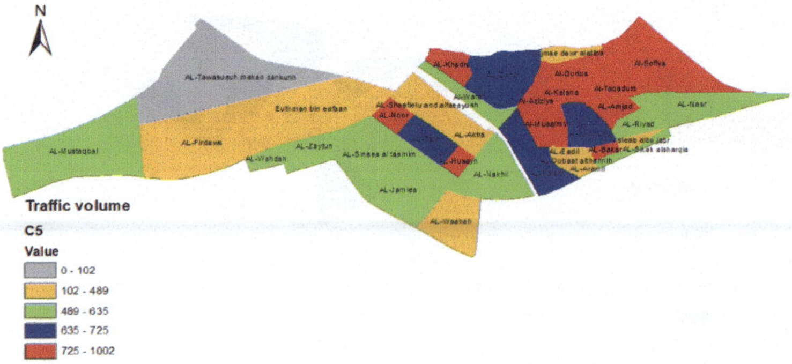

Fig. 10. Traffic volume criteria (C5).

- Green color represented Medium suitability district for parking area.
- Blue color represented high suitability district for parking area.
- Red color represented very high suitability district for parking area.

The raster calculator was used to generate the map that was produced as a consequence. GIS software were utilized in order to identify the locations of the criteria, and density

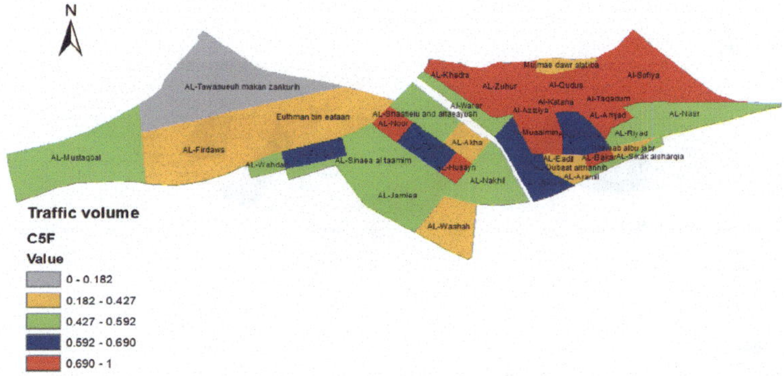

Fig. 11. Traffic volume criteria (CF5).

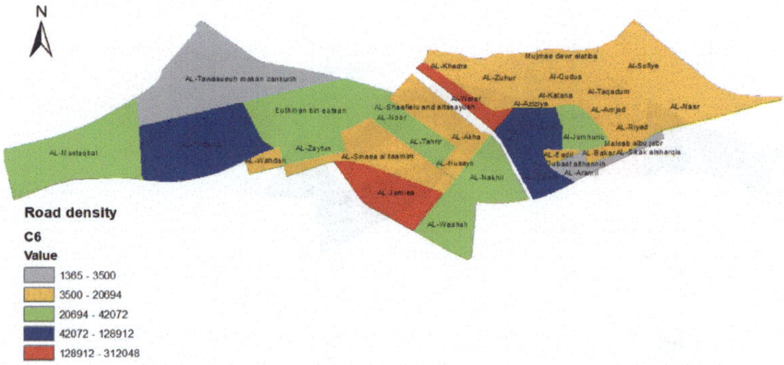

Fig. 12. Road density criteria (C6).

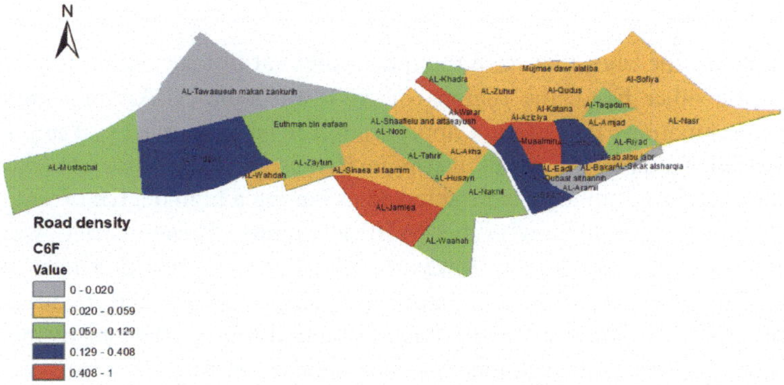

Fig. 13. Road density criteria (CF6).

assessments were carried out for each of the selected criteria. According to the information shown above, the raster data that was obtained were divided into five distinct categories. The maps presented here illustrate the appropriate parking areas in each district of Ramadi city, based on the AHP and FAHP methods (Figs. 14 and 15). According to the findings that are provided in Figs. 14 and 15, the districts of Al-Muaalmin, Al-Warar,

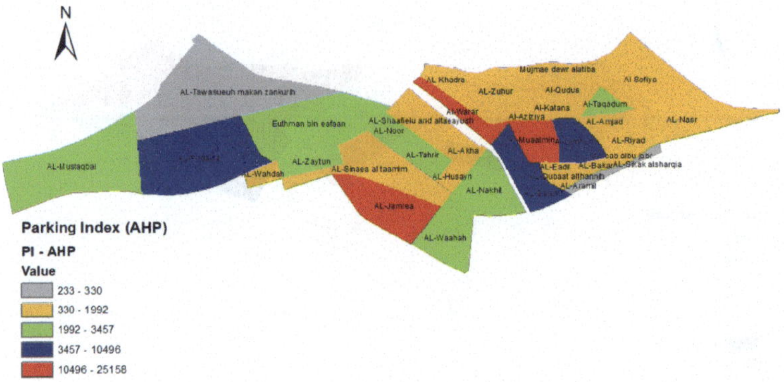

Fig. 14. Parking Index (PI – AHP).

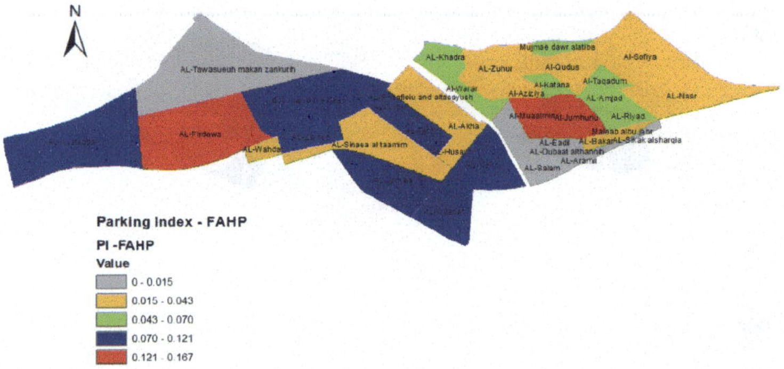

Fig. 15. Parking Index (PI – FAHP).

and Al-Jamie are the ones that have the most potential to take on the role of parking places. On the other hand, the regions of Al-Tawasueuh, Al-Aramil, and Al-Sikak Alsharqia are situations that are not appropriate for parking places of any kind. The aforementioned features are what distinguish these distinct districts. Al-Firdaws, Al-Muaalmin, and Al-Jumhurer were selected as having a high degree of suitability for parking places in accordance with the FAHP technique. These districts were picked because of their level of suitability. However, when it comes to parking areas, the neighborhoods of Al-Salam, Al-Adil, Al-Aramil, Al-Sikak alsharqia, and Al-dubaat althannih are not very good options. The findings presented in Figs. 14 and 15 indicate that the most favorable areas for establishing new parking area for the districts that have large area, many amenities, high land price, high car ownership, high traffic volume, and high road density. These districts demonstrate greater appropriateness ratings in terms of their capability for accommodating parking facilities (see Table 3).

Table 3. Suitability Area for Parking.

Suitability	AHP (km2)	AHP (%)	FAHP (km2)	FAHP (%)
Very High	4.515	8.07%	7.443	13.31%
High	7.911	14.14%	10.167	18.18%
Medium	16.725	29.90%	4.479	8.01%
Low	20.035	35.82%	39.963	71.45%
Very Low	6.748	12.06%	2.882	5.15%

4 Conclusions

The rapid rise of urban populations and the unplanned construction of cities have resulted in the emergence of urbanization concerns and challenges, particularly in the realm of traffic and parking. These issues and challenges will be discussed more below. The approach that is being discussed here has a number of distinct advantages, one of which is that it is designed to make the process of finding appropriate parking locations easier. In order to provide a solution that is both more efficient and effective, this selection process, which makes use of the combined AHP and FAHP approach as well as GIS applications, signifies a difficult decision-making dilemma. In order to ascertain the relative weights of the variables that were predetermined, the AHP and FAHP approach was applied. In particular, the selection of appropriate parking places in thirty-six Ramadi districts based on the impact of specified variables such as the amount of land available, the amenities available, the number of cars owned, the amount of traffic, and the density of the roads. When dealing with site selection issues, one way that is frequently used is the AHP and FAHP method. This method involves evaluating the relevance levels of criteria that influence site selection by using the opinions of specialist individuals. The weights of the preset criteria were computed through the construction of pairwise comparison matrices. Following this, the spatial locations of the criteria were determined by utilizing the GIS applications. As a consequence of this analysis, a map was created that illustrates the places within thirty-six Ramadi districts that are located in an appropriate parking spot. According to the results of the AHP and FAHP approaches, it is possible to draw the conclusion that the factors of area, amenities, land price, automobile ownership, traffic volume, and road density have a weighted effect of 21%, 17%, 16%, 27%, 11%, and 8% respectively on the acceptability of parking areas in the sixty-three districts of Ramadi city. This conclusion can be reached given the findings of the AHP and FAHP approaches. Al-Muaalmin, Al-Warar, and Al-Jamie are the districts that have the highest suitability for parking areas, according to the findings of the AHP technique. Al-Tawasueuh, Al-Aramil, and Al-Sikak Alsharqia are, on the other hand, districts that have a very low suitability for parking areas of any kind. The high suitability district for parking area was Al-Firdaws, Al-Muaalmin, and Al-Jumhurer, according to the FAHP technique. When it comes to parking areas, however, the districts of Al-Salam, Al-Adil, Al-Aramil, Al-Sikak alsharqia, and Al-dubaat althannih are not very suitable.

References

1. Alkan, T., Durduran, S.S.: GIS-supported mapping of suitable parking areas using AHP method: The case of Konya. Int. Arch. Photogramm. Remote. Sens. Spat. Inf. Sci. **46**, 51–56 (2021). https://doi.org/10.5194/isprs-archives-XLVI-4-W5-2021-51-2021
2. Fahim, A., Hasan, M., Chowdhury, M.A.: Smart parking systems: comprehensive review based on various aspects. Heliyon. **7**(5), e07050 (2021). https://doi.org/10.1016/j.heliyon.2021.e07050
3. Farzanmanesh, R., Naeeni, A.G., Abdullah, A.M.: Parking site selection management using fuzzy logic and multi criteria decision making. Environ. Asia. **3**(3), 109–116 (2010)
4. Ozturk, D., Kılıç-Gul, F.: GIS-based multicriteria decision analysis for parking site selection. Kuwait J. Sci. **47**(3), 2–14 (2020)
5. Jelokhani-Niaraki, M., Malczewski, J.: A group multicriteria spatial decision support system for parking site selection problem: a case study. Land Use Policy. **42**, 492–508 (2015). https://doi.org/10.1016/j.landusepol.2014.09.003
6. Aliniai, K., Yarahmadi, A., Zarin, J.Z., Yarahmadi, H., Lak, S.B.: Parking lot site selection: An opening gate towards sustainable GIS-based urban traffic management. J. Indian Soc. Remote Sens. **43**, 801–813 (2015). https://doi.org/10.1007/s12524-014-0415-3
7. Kazazi Darani, S., Akbari Eslami, A., Jabbari, M., Asefi, H.: Parking lot site selection using a fuzzy AHP-TOPSIS framework in Tuyserkan, Iran. J. Urban Plan. Dev. **144**(3), 04018022 (2018). https://doi.org/10.1061/(ASCE)UP.1943-5444.0000456
8. Hensher, D.A., King, J.: Parking demand and responsiveness to supply, pricing and location in the Sydney central business district. Transp. Res. A Policy Pract. **35**(3), 177–196 (2001). https://doi.org/10.1016/S0965-8564(99)00054-3
9. Demir, S., Basaraner, M., Gumus, A.T.: Selection of suitable parking lot sites in megacities: a case study for four districts of Istanbul. Land Use Policy. **111**, 105731 (2021). https://doi.org/10.1016/j.landusepol.2021.105731
10. Neisani Samani, Z., Karimi, M., Alesheikh, A.A.: A novel approach to site selection: collaborative multi-criteria decision making through geo-social network (case study: public parking). ISPRS Int. J. GeoInf. **7**(3), 82 (2018). https://doi.org/10.3390/ijgi7030082
11. Awan, F.M., Saleem, Y., Minerva, R., Crespi, N.: A comparative analysis of machine/deep learning models for parking space availability prediction. Sensors. **20**(1), 322 (2020). https://doi.org/10.3390/s20010322
12. Feng, Y., Xu, Y., Hu, Q., Krishnamoorthy, S., Tang, Z.: Predicting vacant parking space availability zone-wisely: a hybrid deep learning approach. Complex Intell. Syst. **8**(5), 4145–4161 (2022). https://doi.org/10.1007/s40747-022-00700-1
13. Benenson, I., Martens, K., Birfir, S.: PARKAGENT: an agent-based model of parking in the city. Comput. Environ. Urban. Syst. **32**(6), 431–439 (2008). https://doi.org/10.1016/j.compenvurbsys.2008.09.011
14. Tian, D., Wang, Y., Lu, G., Yu, G., Duan, X., Liu, H.: A parking space finding method based on VANETs. In: ICCTP 2011: Towards Sustainable Transportation Systems, pp. 1511–1520 (2011). https://doi.org/10.1061/41186(421)150
15. Hosseini Lagha, G., Mokhtary Malek Abadi, R., Gandomkar, A.: Geographical analysis of parking land use in Genaveh applying AHP Model. J. Urban Reg. Stud. Res. **4**(13), 95–114 (2012)
16. Shoup, D.C.: Cruising for parking. Transp. Policy. **13**(6), 479–486 (2006). https://doi.org/10.1016/j.tranpol.2006.05.005
17. Shoup, D.: Free parking or free markets. In: Parking and the City, pp. 270–275. Routledge (2018)

18. Giuffrè, T., Siniscalchi, S.M., Tesoriere, G.: A novel architecture of parking management for smart cities. Procedia Soc. Behav. Sci. **53**, 16–28 (2012). https://doi.org/10.1016/j.sbspro.2012.09.856
19. Caicedo, F.: Real-time parking information management to reduce search time, vehicle displacement and emissions. Transp. Res. Part D: Transp. Environ. **15**(4), 228–234 (2010). https://doi.org/10.1016/j.trd.2010.02.008
20. Gülhan, G., Ceylan, H.: Otopark sorununa otopark yönetimi temelinde yaklaşımlar: İzmir Örneği. Dokuz. Eylül Üniv. Mühendis. Fak. Fen. Mühendis. Derg. **12**(1), 63–73 (2010)
21. Saaty, T.L.: The Analytic Hierarchy Process. McGraw-Hill, New York (1980)
22. Guo, S., Zhao, H.: Optimal site selection of electric vehicle charging station by using fuzzy TOPSIS based on sustainability perspective. Appl. Energy. **158**, 390–402 (2015)

Community Charging of Electric Cars in Residential Areas with Multi-Story Buildings

Sebastian Naumann[1] (✉), Julia Brandt[2], Kaja Cordes[3], and Elmar Lukas[3]

[1] Institut f. Automation und Kommunikation e.V., Magdeburg, Germany
sebastian.naumann@ifak.eu

[2] Die Stadtfelder Wohnungsgenossenschaft eG, Magdeburg, Germany
julia.brandt@diestadtfelder.de

[3] Otto-von-Guericke University, Magdeburg, Germany
kaja.cordes@ovgu.de; elmar.lukas@ovgu.de

Abstract. In Germany, the share of the greenhouse gas emissions from the transport sector has remained constant at around 20% for years. However, according to the German Climate Protection Act, it should decrease to zero until 2045. In order to achieve this goal, replacing combustion engine cars by battery electric cars is an important pillar. In Germany, however, about 50% of households live in multi-story buildings that currently lack sufficient charging infrastructure. While single and two-family houses can be easily upgraded with a wallbox off the shelf, the power grids connecting to multi-story buildings are not designed to manage the simultaneous charging of batteries in larger quantities. We present an efficient and cost-effective approach to coordinate the charging actions to share limited charging power by means of a smartphone app to overcome this problem. Furthermore, the paper quantifies the benefits of planning the charging actions according to the weather forecast for maximal exploiting any available solar power.

Keywords: Battery electric vehicle (BEV) · Weather forecast · Solar power · Residential areas · Multi-story buildings · Community charging · Shared charging

1 Introduction

According to the German Climate Protection Act, Germany's greenhouse gas emissions should be reduced by 65% relative to 1990 levels by 2030 [1]. The greenhouse gas emissions are divided into six sectors: agriculture, transport, buildings, industry, energy and waste. In 2022, the transport sector's share was 19,8%, which corresponds to 147.9 million t of CO2 equivalents (see Fig. 1). Unlike other sectors, the greenhouse gas emissions of the transport sector, however, have remained at this level for years.

In order to reach the greenhouse gas reduction targets the transport sector must continuously decrease its emissions until reaching zero by 2045. In 2023, the emissions were at 146 million t, which did not match the target of 133 million t (see Fig. 2).

G. Sierpiński et al. (Eds.): TSTP 2024, LNNS 1257, pp. 77–91, 2024.
https://doi.org/10.1007/978-3-031-81777-9_6

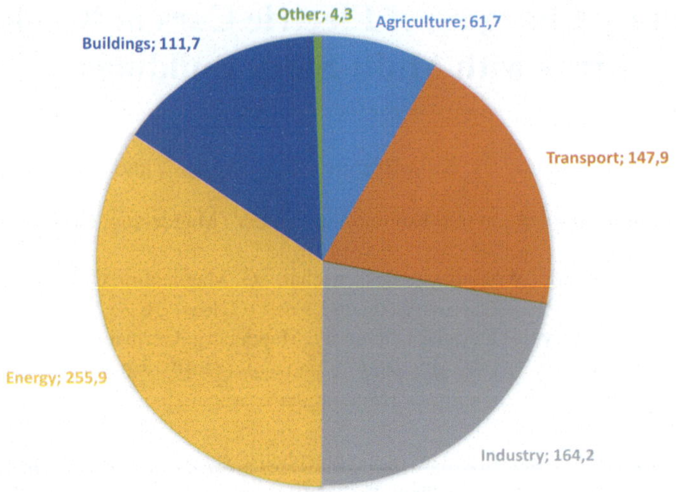

Fig. 1. Greenhouse gas emissions in Germany by sector 2022; Amount in millions of tons CO2 equivalents [2].

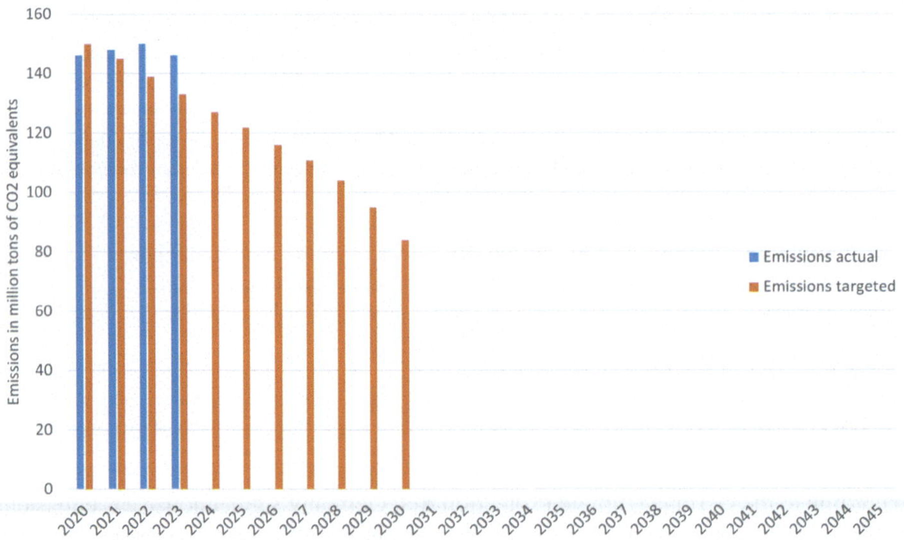

Fig. 2. Development of greenhouse gas emissions in the transport sector in million tons CO2 equivalents in Germany [3].

The German Climate Protection Act from 2021 [1] states that each sector has to reduce its emissions; if not, then immediate measures, such as unpopular restrictions on driving on certain days in the transport sector, must be implemented. With the amendment from 2024 [4] this requirement has been softened. If the transport sector fails to meet its targets and another sector is able to compensate the deficit, no actions must be taken.

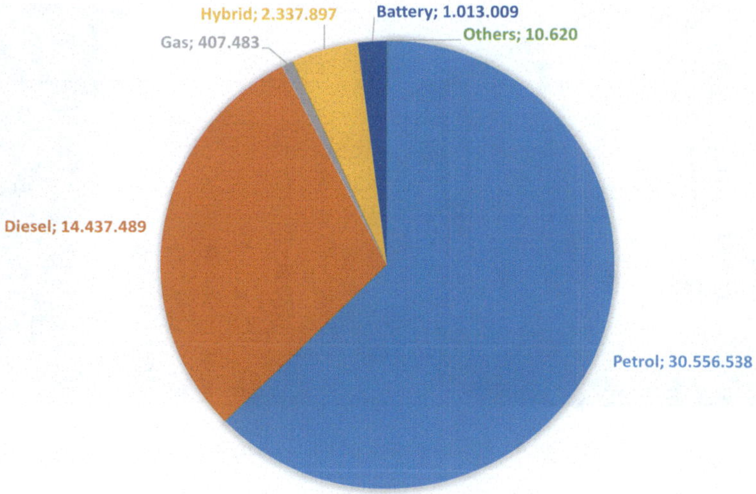

Fig. 3. Stock of cars in Germany [5].

In Germany, the number of registered cars is approximately 48 million (48,763,036 as of 1.1.2023). Figure 3 reveals the distribution of the cars according to their drive type [5]. The aim is to reach 15 million fully electric cars by 2030. According to the European Regulation 2023/851 rules, from 2035 onwards no new cars with combustion engine are newly registered [6].

Charging the car is an important prerequisite for users considering purchasing a battery electric vehicle (BEV). While residents of single or two-family households with permanent parking places can easily install a wallbox, approximately 50% of households in Germany live in urban multi-story buildings which currently massively lack sufficient charging infrastructure. The reasons are multi-layered and are described in detail in Sect. 2. Sections 3–5 then present an approach for overcoming the issue of charging electric cars in residential areas with multi-story buildings by community charging.

2 Requirements and Prerequisites

2.1 Multi-Story Apartment Houses

Multi-story apartment houses are widespread in urban areas all over the world. In such environments permanent parking places for users are seldom. The most important prerequisite for people of buying a BEV is to have a charging opportunity at home. Figure 4 shows the residential area of the Stadtfelder housing cooperative in Magdeburg, Germany, being addressed within this paper.

There are buildings with four or more floors, some garages and some parking slots for the tenants whereas mostly they are not destinated for certain users. Furthermore,

Fig. 4. Environment of BEV charging.

Fig. 4 also shows three cars at the front right, which are currently being charged at charging stations installed within the residential area about 2 years ago.

2.2 BEV Ownership Survey

Within the framework of the project, a survey was conducted by the Otto-von-Guericke University Magdeburg among the tenants of the Stadtfelder housing cooperative. A call for participation was distributed to approximately 4800 tenants, and 120 of these tenants completed the survey in full. Of the respondents, about 53% are male, 42% female, and 2% non-binary, with an average age of 43.28 years. Regarding their employment status, 58% of the tenants stated that they are employed, while approximately 18% each are retired or students.

The following analysis focuses solely on the evaluation of questions concerning mobility behavior, e-mobility, and the acceptance of apps for controlling and planning charging processes. When tenants were asked about their basic mobility behavior 40% indicated that the car is the most frequently used mode of transportation, followed by public transportation and bicycles as shown in Fig. 5.

During the week, tenants report driving an average of 100 km per day, while on weekends, the distance drops to 50 km per day. Approximately 65% of tenants own a car, of which just over 88% are powered by gasoline or diesel (see Fig. 6). Only about 4% of these cars are fully electric (see Fig. 6). Furthermore, the survey indicates that a quarter of the tenants have already had experience with fully electric vehicles. However, the majority of tenants have not yet had any experience with alternative engines (fully electric, hybrid, or plug-in hybrid).

To estimate the potential demand for charging stations over the next 2 years, tenants were asked about the likelihood of purchasing an electric vehicle within this period. Currently, 14% of tenants are considering this option. About 43% cited improvements in charging infrastructure as a prerequisite for purchasing an electric vehicle, and approximately 33% mentioned a reduction in acquisition costs. Additionally, shorter

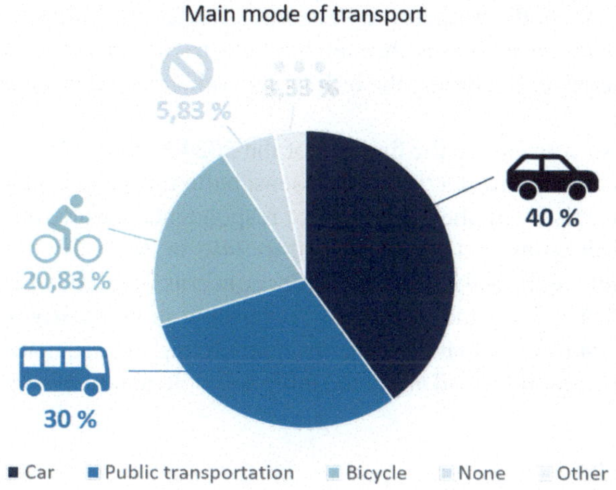

Fig. 5. Overview of main mode of transport.

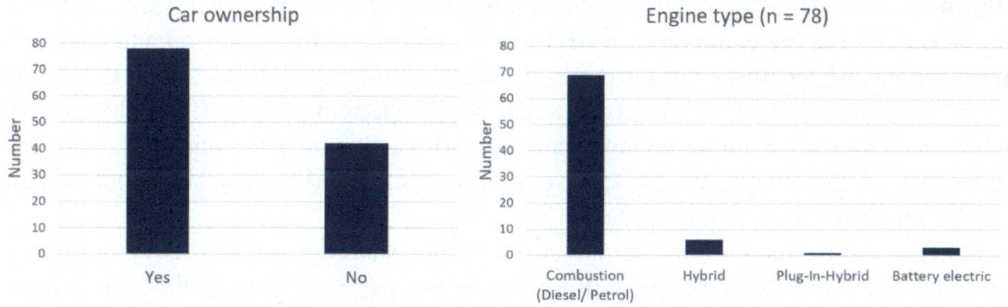

Fig. 6. Overview of car ownership and car type specifications.

charging times and longer driving ranges are factors that influence tenants' decisions regarding such a purchase. Concerns among tenants include the charging process, environmental impact, battery life, and safety. Tenants were further asked whether they have external charging options, such as at work or when shopping, as this also influences the demand for a charging point. Around 47% stated that they also have the opportunity to charge an electric car externally.

Another section of the questionnaire addresses questions regarding the charging management system and the handling of charging processes. Two different charging management systems, presented in the form of an app, were compared. The main difference between the two systems lies in the allocation of charging slots. In one app, charging demands are optimized based on an algorithm, which then assigns charging slots to tenants. The other app resembles a reservation management system, where each tenant can select and reserve a charging slot as long as it is still available. Two-thirds of the tenants indicated a preference for the reservation management app. This implies that the tenants have a preference towards solutions that are transparent and offer the highest amount of individual flexibility. Hence, proprietary solutions that have black-box character that follow a top down approach do not yield much acceptance. As for the preferred billing method for charging sessions, 40% chose billing through the app, while nearly 20% preferred billing through their electricity provider, and the remaining 40% had no preference. Regarding the billing interval, 45% preferred billing after each charging session.

The results also corroborate the findings of the 2024 Mobility Monitor published by acatech Deutsche Akademie der Technikwissenschaften. It reveals that concerns about electromobility persist, with the proportion of respondents considering the purchase of an electric car decreasing to 17%, compared to 24% in 2020. The reservations about electric cars largely align with our results. Key concerns highlighted include the range of electric vehicles and whether they truly represent a more environmentally friendly alternative. The data shows that respondents considering the acquisition of an electric car are generally better informed about the topic and hold a more positive view of it [7].

2.3 User Groups

There are two user groups being addressed:

Organizations The challenges for the employees of the Stadtfelder housing cooperative when sharing the cooperative's BEVs are complex: there are several and changing users for the BEVs, there are currently 3 electric charging stations for 6 BEVs, and despite illness, vacation and appointments, all BEVs should be sufficiently charged before each journey without having a caretaker who acts as a "gas station attendant and dispatcher". Furthermore, the cheaper solar power generated by their own photovoltaic system should be used to most extend.

When a group (the employees) has to look after shared resources (BEVs), there is always the risk that some people will become more committed than others, leading to dissatisfaction that contributes to a negative team culture overall. A lack of clarity about responsibilities is a common cause, but social influences can also play a role. It is important that the group is aware of these dynamics and takes steps to ensure that the work is distributed fairly and that all members can contribute equally [8].

Individuals Private individuals are a further user group. Compared to organizations they do not share common BEVs.

2.4 Power Grid

The connection to the power grid of the considered building has only a limited capacity. It was designed decades ago for supplying household appliances like lights, washing machines, dish washers and electrical cooking stoves. It was not designed for many people to charge their BEVs. Following [9], only a small amount of BEVs could be charged simultaneously with a power of 11.0 kW. A higher power demand would damage the power cable and the transformer. Therefore, one option for enabling BEV charging at residential areas is to upgrade the power network components which would be an expensive and time-consuming task. Another option is, to coordinate the charging operations within the community. By this, the existing power components can be used by a lot of people, and no further building measures are necessary. This approach is being addressed within this paper.

2.5 Photovoltaic System

The Stadtfelder housing cooperative owns a photovoltaic system. The German Renewables Energy Act (EEG) [10] created in 2001 was established to help to achieve a market-breakthrough of renewable energies like wind and photovoltaic power systems. The core of the EEG is that owners of a photovoltaic system or other systems for generating electricity from renewable energies that are connected to the electricity grid are guaranteed a fixed price for each feed-in to the electricity grid. The EEG has been amended several times in recent years, primarily to adapt the feed-in tariff to increasing economic efficiency.

Meanwhile, electricity from renewable resources has become very cheap. Hence, the feed-in tariffs have been decreased, such that the most economical way is to consume the generated electricity directly, e.g. by charging the BEV [11, 12].

2.6 Legal Issues

German National Building Electromobility Infrastructure Act National legislation has established conditions and scenarios that necessitate the installation of charging stations in residential areas. The goal is to provide sufficient infrastructure to support the adoption of electric vehicles in Germany in the coming years. The Building Electromobility Infrastructure Act (GEIG) [13] distinguishes between existing buildings and new buildings and then again between residential and non-residential buildings. For residential buildings, the GEIG Act stipulates the provision of charging infrastructure for existing or newly constructed parking spaces on the property or in the vicinity of the respective building. For new residential buildings with more than 5 parking spaces, each parking space must be equipped with the necessary electrical infrastructure to prepare them for the installation of charging stations if required. In the case of a renovation of more than 25% of the external envelope of an existing building that has more than 10 parking spaces on the property or in the vicinity of the building, the electrical infrastructure must be prepared for each parking space, even in the case of a general renovation of the electrical infrastructure of the building or the parking space itself.

European Energy Performance Building Directive The Energy Performance Building Directive EU (EU) 2024/1275 that has been completely recasted in 2024 [14], sets new requirements that must be implemented into national legislation for charging infrastructure in residential areas. Concerning new housing buildings, 50% of the parking spaces need to be pre-cabled and ducted and additionally at least one charging point must be installed if there are more than three parking spaces. For major renovations on buildings with more than three parking spaces in or close to the housing building, pre-cabling and ducting of at least 50% of the parking spaces must take place. Therefore, national legislation must be adopted in the coming next 2 years. Housing companies must fulfil those obligations to provide infrastructure for charging electrical vehicles at home.

German Measurement and Calibration Act In addition to these regulations for the building itself, another question arises. The fairest solution is for users to pay exactly for the electricity they charge. However, this requires a calibrated electricity meter, as the charging of BEVs is subject to the German Measurement and Calibration Act [15], which regulates the provision, use and calibration of measuring devices such as water or heat meters. The law protects consumers and fair trading. Furthermore, the cable between the electricity meter and the plug must not extend a length of 5.0 meters. Unfortunately, this implies for the operators of charging stations: A calibrated electricity meter is needed, and the electricity meter has to be recalibrated or exchanged every few years. This adds another cost and time factor and makes operating a charging station more expensive. One way of circumventing the regulation is to sell the electricity for charging at a flat rate. However, a flat-rate tariff harbours a risk for both the operator and the customer. Depending on how the electricity price develops, it can happen that the flat-rate price is set too low or too high, meaning that the operator does not receive enough money or the customer pays more than necessary. Both sides would have to agree on a price for such a flat rate. In this case, the operator does not have to use a calibrated electricity meter or adhere to any official procedures and therefore has lower costs to pass on to the customer.

3 Approach

3.1 Shared Charging Calendar

The general approach of community charging, presented in this paper, is to set up a few charging stations (currently between three and six) that are used by all residents of the surrounding residential buildings, as well as organisations like Stadtfelder housing cooperative as mentioned in Sect. 2.2. A shared calendar is used to book time slots for charging the cars. In order to optimally exploit the available charging stations, the concrete charging stations are assigned to the booked charging time slots by the booking system using a Greedy algorithm as described in [16]. If the choice of the charging station were left to the users, unfavorable combinations of charging time slots could otherwise occur, blocking further coherent bookings. The technical aspects of this approach are described in more detail in Sects. 4 and 5.

3.2 Alternative Approaches

Booking charging time slots and driving the car to the charging station for plugging in and out as well as moving the car away, is an inconvenient task that must fit into the user's regular daily routine. A more convenient way could be controlled charging. For this, the car to be charged would be connected to the charging station for a period of time that exceeds the required time for charging. (see Fig. 7) This makes it possible to change the charging power during the charging process depending on external circumstances (e.g. availability of solar power).

Compared to the shared calendar, the approach of controlled charging offers some advantages as better exploiting renewable energies, better reacting on power grid restrictions as well as enabling bidirectional charging. On the downside, much more

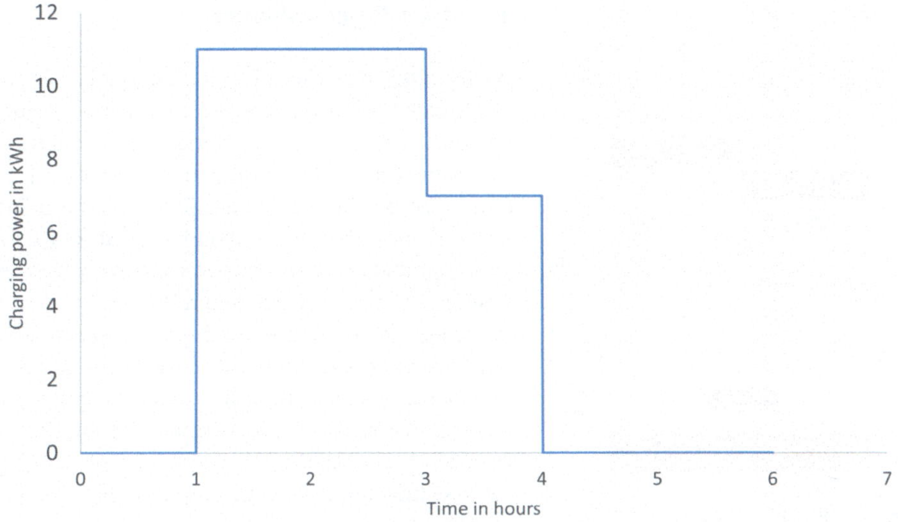

Fig. 7. Controlled charging.

charging stations are needed which leads to higher investment costs that might not be economically reasonable. However, if we think of only one core charging station with a lot of simple and cheap sockets, the situation would be quite different. In this context, a legal question arises. If someone would like to sell energy per kWh (billing with precise charged amount of energy in kWh), an electricity meter is that complies with calibration law is required. Additionally, the cable to the car plug must be no longer than 5 meters. A solution could be billing with a flat rate, meaning users pay a fixed amount each month regardless of the amount of energy they charge. In this case, a calibrated electricity meter is not required, and there is no need to adhere to a maximum cable length. The question for the owner is how to set the flat rate in order to avoid significant losses while remaining attractive to users. However, this is a theoretically consideration only and is not further considered within this paper.

4 Smartphone App

4.1 Booking Charging Time Slots

As mentioned earlier, a smartphone app has been developed to coordinate the shared use of the charging stations. Figure 8 shows the app view for booking a charging time slot. The symbol depicting sunlight on the solar panels signifies a predicted high solar irradiation value, suggesting that the self-generated energy from the photovoltaic system is likely to be available for charging. The solar radiation forecast values are provided by the German Weather Service.

Fig. 8. Booking a charging time slot.

4.2 User Responsibilities

As introduced in Sect. 2.3, the employees of the Stadtfelder housing cooperative form a user group. Together they are responsible for charging the batteries of the cooperative's BEVs. If an employee has booked a charging time, he is initially responsible for moving the BEV to the charging station at the respective time, plugging it in or out after charging and releasing the charging zone again. If the employee is unavailable due to reasons such as holiday, illness or appointments, they can use the app to request a colleague to drive the BEV to the charging station, plug it in or unplug it and drive it away. A corresponding request then appears in the app (see Fig. 9). The employees who see this request have the option of agreeing to it. If another employee accepts the request, the sender also receives a notification on their smartphone. Additionally, the responsible employee receives a reminder on their smartphone 15 min prior to the start and end of the charging process.

4.3 Backend

The smartphone app of course needs a backend e.g. a database with some tables storing the data. We use a WordPress (https://wordpress.com) installation coming with a MySQL database (https://www.mysql.com) and a user management which are very good prerequisites for our needs.

4.4 Limitations

Currently, there is no link between the presented app and the charging stations. So, one could use the charging station without having booked a charging time slot. In practical use, this may cause social problems and is subject to observation.

4.5 Evaluation

The smartphone app has been tested by the employees at the Stadtfelder housing cooperative for some weeks. It has proven to be a smart and efficient way to organize the charging of the cooperative's BEVs by prioritizing times with high solar radiation. The data collected on charging times and BEVs charged provides a lot of useful information on how to further improve BEV charging from an organizational point of view.

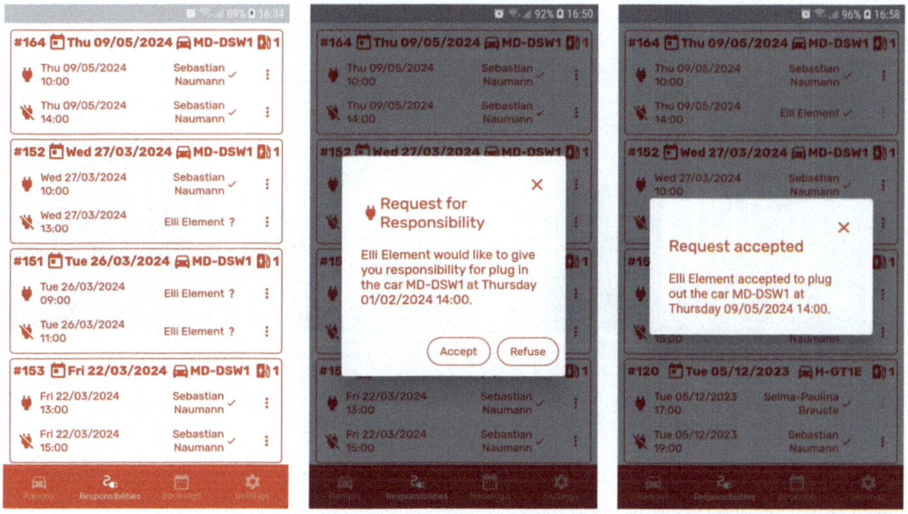

Fig. 9. Delegation of responsibility.

5 Solar Exploitation

5.1 Available Solar Power

As described in Sect. 2.5, there is a financial difference between consuming the energy generated by the photovoltaic system directly and consuming the energy from the grid. Currently, the Stadtfelder housing cooperative states the self-usage costs at 0.3451 €/kWh and the grid costs at 0.4598 €/kWh. As described in Sect. 4.1, we use the weather forecast to determine the times a lot of self-generated solar power is likely to be available. Figure 10 shows the power provided by the photovoltaic power plant between 2023/Sep/3 and 2024/Mar/31 in kW. It can be seen that days with high solar irradiation values alternate with days with very low solar irradiation values. Accordingly, it should be advantageous to book charging times based on the forecast of high solar irradiation values in order to maximize the availability of solar energy. In order to quantify the benefits, we have investigated this time period in more detail.

5.2 Assumptions

First of all, a base load of 4.5 kW is assumed, which is always consumed by the various electrical devices in the building. This base load is always subtracted from the solar power generated. Second, it is assumed that the Stadtfelder housing cooperative operates a total of 6 BEVs. Four BEVs of the type Volkswagen ID.3 and two BEVs of the type Volkswagen ID.4. According to the General German Automobile Club (https://www.adac.de), the energy consumption of the ID.3 is 19.2 kWh/100 km, while that of the ID.4 is 22.8 kWh/100 km. An energy requirement of 100 km/week is presumed for each BEV, resulting in a total consumption of 122.4 kWh/week. An ID.3 is charged

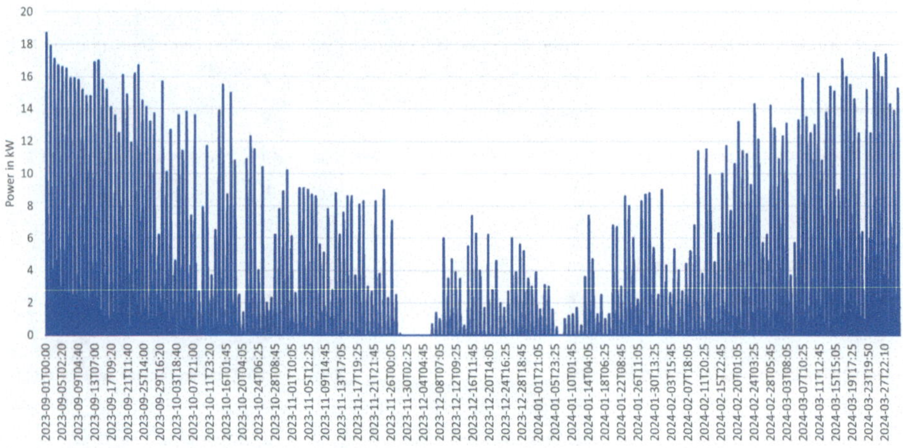

Fig. 10. Power generated by the installed photovoltaic system from 2023/Sep/3 to 2024/Mar/31 in kW with a resolution of 5 min.

with a power of 7.4 kW whereas an ID.4 is charged with 11.0 kW. Furthermore, it is assumed that three charging stations are available and that there are regular office hours (Monday to Friday from 8 am to 5 pm) during which a BEV can be connected to a charging station.

5.3 Usage Variants

We consider three types of use that have an impact on the exploitation rate of solar power and are explained in more detail below.

WOWF (Without Weather Forecast) This usage variant assumes not to follow any predicted favorable charging conditions. Instead, users will charge the BEVs at any time. In order to reproduce this behavior, we distribute each BEV charging action among the mentioned office times randomly and calculate the exploited solar energy value by comparing it with the solar power available. In order the get a usable value, we repeated this 100.000 times and calculated the average.

WWF (With Weather Forecast) An algorithm in order to select the optimal charging times over all cars is not easy to implement. Therefore, in this project a simple best-first algorithm is implemented. First, the BEV charging requests are sorted according to their charging power, with the highest value in first place. Then each BEV is searched for the period with the highest solar power, so that the BEV can be charged predominantly with solar power. This means that if a BEV requires 11.0 kW for 2 h, for example, the best charging period would be the one in which the photovoltaic system delivers 11.0 kW or more (excluding the base load). A period in which the first hour delivers 22.0 kW and the second hour 0.0 kW would result in 11.0 kWh and not 22.0 kWh, as the BEV can only be charged with a maximum of 11 kWh. When the best period has been found, the charging process is assigned to this period and the available solar power for this period is reduced by the charging power (11.0 kW in this case).

MAX This usage variant exploits all available solar power. All charging time and charging power is based on it. Currently, it is a theoretically variant only. For example, if there is a solar power of 10.0 kW available, an ID.4 would be charged with 10.0 kW only instead of 11.0 kW as in the WOWF and WWF usage variants, and the charging time would be extended accordingly.

5.4 Results

Table 1 shows the results obtained for the period from 2023/Sep/3 to 2024/Mar/31 as a whole. The column Q_{Solar} shows the solar power self-usage rate of the usage variant. S_{Costs} corresponds to the financial savings of the usage variant compared to the total costs C_{Total} of usage variant WOWF. While Q_{Solar} is 9.6% for WOWF, WWF increases this value to 29.6%. The financial savings in this period amount to 73.7 €.

Figure 11 shows the results obtained per week. The lines show the solar energy exploited from the installed photovoltaic system depending on the usage variant.

Table 1. Impact of the usage variant on the solar power self-usage rate.

Solar usage	E_{Solar} *kWh*	E_{Grid} *kWh*	Q_{Solar} %	C_{Solar} €	C_{Grid} €	C_{Total} €	S_{Solar} *kWh*	S_{Costs} €	S_{Costs} %
WOWF	360.1	3383.9	9.6	124.3	1556.0	1680.2	–	–	–
WWF	1002.2	2741.8	29.6	345.9	1260.7	1606.6	642.1	73.7	4.4
MAX	2552.8	1191.2	75.0	881.0	547.7	1428.7	2192.7	251.5	15

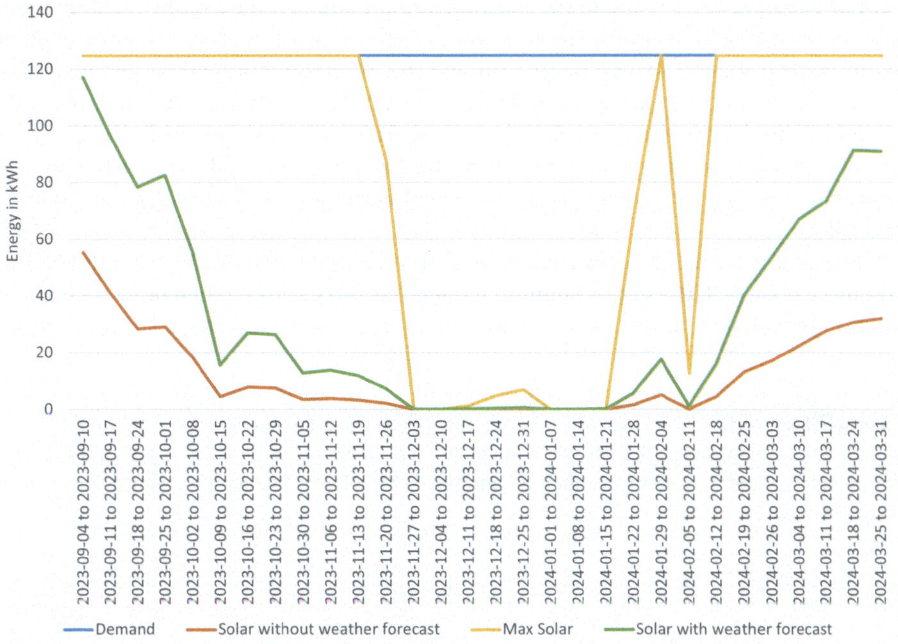

Fig. 11. Solar power usage depending on user behavior from 4/Sep/2023 until 31/Mar/2024.

In September and October as well as February and March, the difference of E_{Solar} between WOWF and WWF is maximal, so an increasing Q_{Solar} and increasing cost savings S_{Costs} are expected when the period from April to August 2024 is additionally considered during further investigations.

6 Conclusions

In Germany, about 50% of households live in multi-story buildings that currently lack sufficient charging infrastructure for BEVs. With this paper, we presented an efficient, cost-effective and quickly implementable approach to coordinate the charging actions under the restrictions of limited charging power and a limited number of charging stations. A smartphone app is used to book charging time slots in a shared calendar. The option of responsibility transfer helps organizations charging its shared BEVs. It also promotes a positive team culture by implementing clear responsibilities and clear communication. Overall, the app offers an answer to the specific requirements of organizations and makes it easier to integrate an environmentally friendly fleet into everyday business life. Finally, the paper quantifies the benefits of planning the charging actions according to the weather forecast for maximal exploiting any available solar power directly. How the employees of the Stadtfelder housing cooperative can be persuaded to book exactly these times for charging, and to what extent automatic selection is possible, will be investigated in the next months.

From our experience, there is a very small amount of people behaving not cooperative. For example, they could book many charging points without really needing them, block the charging station for longer than booked or block a charging point without having booked it at all. These people run the risk of creating a general dissatisfaction with the system and among the participants through their behaviour. As mentioned, the amount of those egocentric people is very low. However, it seems to be necessary to introduce mechanisms to prevent uncooperative behavior. One of these is to link the app to the charging stations so that nobody can use a charging station without having booked a time slot.

The gap of solar radiation in winter could be very well expanded by a combined heat and power plant (CHP): in late autumn, winter and early spring when not enough solar radiation is available, more heat is needed and power produced by the CHP for charging the cars.

Acknowledgements. The work presented in this paper has been carried out within the German national research project ELEMENT funded by the German Federal Ministry of Economic Affairs and Climate Action under the grant number 01ME21003. The authors would like to thank for the financial support.

References

1. German Climate Protection Act, 24. Juni 2021. https://www.gesetze-im-internet.de/ksg/KSG.pdf
2. Renewable Energy Agency: Greenhouse gas emissions in Germany by sector 2022. https://www.unendlich-viel-energie.de/mediathek/grafiken/treibhausgasemissionen-in-deutschland-nach-sektoren-2022
3. German Federal Environment Agency: Climate protection goals in the transport sector (2024). https://www.umweltbundesamt.de/themen/verkehr/klimaschutz-im-verkehr#ziele
4. German Climate Protection Act Amendment 2024. 20/8290 - Gesetzentwurf: Entwurf eines Zweiten Gesetzes zur Änderung des Bundes-Klimaschutzgesetzes. https://dserver.bundestag.de/btd/20/082/2008290.pdf
5. German Federal Motor Transport Authority: Vehicle inventory (press release). (2023) https://www.kba.de
6. Regulation (EU) 2023/851 of the European Parliament and of the Council of 19 April 2023 amending Regulation (EU) 2019/631 as regards strengthening the CO2 emission performance standards for new passenger cars and new light commercial vehicles in line with the Union's increased climate ambition. https://eur-lex.europa.eu/legal-content/EN/TXT/?uri=CELEX:32023R0851
7. acatech – German Academy of Engineering Sciences: Mobility Monitor 2024: Population critically assesses infrastructure and electromobility. (2024, April 29) https://www.acatech.de/allgemein/umfrage-mobilitaetsmonitor-2024/
8. Behfar, K.J., Peterson, R.S., Mannix, E.A., Trochim, W.M.: The critical role of conflict resolution in teams: a close look at the links between conflict type, conflict management strategies, and team outcomes. J. Appl. Psychol. **93**(1), 170 (2008)
9. Massimi, C.: Integration of Charging Infrastructure into the Power Grid Interaction between Energy Management and Network Operation, Master's thesis. Otto von Guericke University Magdeburg, Magdeburg (2023)
10. German Renewable Energy Act. 2000. https://www.bgbl.de/xaver/bgbl/start.xav?startbk=Bundesanzeiger_BGBl&start=//*%5b@attr_id=%27bgbl100s0305.pdf%27%5d
11. Bhatti, A.R., Salam, Z., Aziz, M.J.B.A., Yee, K.P.: A critical review of electric vehicle charging using solar photovoltaic. Int. J. Energy Res. **40**(4), 439–461 (2016)
12. Valerius, G.: Photovoltaics: grid feed-in vs. self-consumption. In: Solar systems guide (2024) https://www.solaranlage-ratgeber.de/photovoltaik/photovoltaik-planung/eigenverbrauch-von-solarstrom
13. German National Building Electromobility Infrastructure Act, March 18th 2021, BGBl. I Page 354
14. DIRECTIVE (EU) 2024/1275 OF THE EUROPEAN PARLIAMENT AND OF THE COUNCIL of 24 April 2024 on the energy performance of buildings (recast), May 8th 2024, Official Journal of the European Union L 2024/1275
15. German Measurement and Calibration Law. https://www.gesetze-im-internet.de/messeg/MessEG.pdf
16. Lenzner, P., Friedrich, T.: Greedy Algorithms. https://hpi.de/friedrich/teaching/units/greedy.html

Electromagnetic Field Emissions by Electric Vehicles

Wioletta Cebulska[1] (✉), Damian Hadryś[1], and Iga Przytuła[1]

[1] Silesian University of Technology, Faculty of Transport and Aviation Engineering, Katowice, Poland
wioletta.cebulska@polsl.pl; iga.przytula@polsl.pl

Abstract. This article discusses the theory of electromagnetic field emissions, the impact on safety, and measures EMF emissions. The electromagnetic field emission of the Dacia Spring electric vehicle was measured in various driving conditions, i.e. in city and highway traffic and during charging. Values such as electric field strength, magnetic field strength and power density were measured. Measurements have shown that the emission of electric field, magnetic field and power density increases with the engine load. The highest electric field intensity recorded during charging was 5 V/m. Compliance with the Regulation of the Minister of Health (2020) indicates that emissions do not exceed permissible levels, ensuring safety. However, temporary symptoms may occur in the immediate vicinity of the vehicle. Continuous monitoring and establishment of exposure guidelines are necessary to reduce the potential health risks associated with electromagnetic fields.

Keywords: Electromagnetic field · Electric vehicle · Health impact

1 Introduction

Currently, significant emphasis is placed on issues related to the protection of the natural environment, especially in the context of the impact of transportation on this environment. The limited amount of mineral resources, climate change, and global warming prompt the automotive industry to adopt more efficient and sustainable solutions. These challenges drive car manufacturers to embrace new technologies and alternative propulsion systems. Examples of such vehicles include electric vehicles (EV) and hybrid vehicles (HEV or PHEV). The issue of electromagnetic field emissions from these means of transport is of importance. The increasing number of vehicles with alternative propulsion systems prompts the discussion of this topic. There is clear evidence that the emission of electromagnetic fields is higher in the case of electric vehicles. Due to certain uncertainties in this area, it seems necessary to examine the issue of emissions originating from the use of automotive transportation. Electromagnetic fields, depending on their intensity and frequency, can cause various ailments [1].

G. Sierpiński et al. (Eds.): TSTP 2024, LNNS 1257, pp. 93–105, 2024.
https://doi.org/10.1007/978-3-031-81777-9_7

2 Physical Interpretation of Electric Field, Magnetic Field, Magnetic Induction, and Power Density

An electric field is a field produced in the vicinity of a charged conductor. The intensity of the electric field represents the force that the electric field exerts on a unit charge, expressed in volts per meter (V/m). On the other hand, a magnetic field is the state of space where forces act on moving electric charges and on bodies with a magnetic moment, regardless of their motion. The intensity of the magnetic field is repressed in amperes per meter (A/m) [2]. The intensities of the electric and magnetic fields are represented by formulas (1) and (2).

$$\mathrm{H} = f(I) \tag{1}$$

$$\mathrm{E} = f(U) \tag{2}$$

I—current intensity, A,
U—voltage, V,
E—electric field intensity, V/m,
H—magnetic field intensity, A/m.

The magnetic field, along with the electric field, is a manifestation of the electromagnetic field with frequencies ranging from 0 Hz (static fields) to 300 GHz (microwave radiation). Above the upper limit of this range, the following types of radiation are distinguished: infrared, visible, ultraviolet, X-ray, gamma, and cosmic radiation. Electromagnetic fields are divided into natural and artificially generated fields. Natural fields are associated with phenomena occurring in the atmosphere (e.g., electrical discharges during a storm), while artificially generated fields are a result of human activity. These fields may differ in wavelength and frequency. However, a common characteristic is their propagation speed in a vacuum [3, 4].

Magnetic induction is a fundamental quantity describing the magnetic field. It describes the intensity of the magnetic field within a body. The unit of magnetic induction is microtesla [μT] or milligauss [mG].

Magnetic induction is represented by formula (3).

$$\mathrm{B} = \mu\mathrm{H} \tag{3}$$

B—magnetic induction, μT,
μ—magnetic permeability, T*m/A,
H—magnetic field intensity, A/m.

Power density is the power per unit area in relation to the direction of radiation propagation. It is expressed in watts per square meter [W/m^2]. Power density S is a measure calculated by a meter based on the measured field intensity values, usually the electric field E, according to the relationship [3–5]:

$$S = \frac{E^2}{377} \tag{4}$$

3 Types of Electromagnetic Field Frequencies

Low frequency (LF) refers to time-varying electric and magnetic fields of low frequency, covering the frequency range from 1 Hz to 100 kHz. LF fields have two components: the electric field caused by an electric charge and the associated magnetic field. Magnetic fields occur only when an electric current flows. The electric component is measured in volts per meter (V/m). The magnetic component is measured in amperes per meter (A/m) and expressed as flux density in teslas (T) or, in some countries, in gauss (G). LF fields are primarily associated with the supply of electrical energy through the generation, distribution, and use of alternating current (AC). The frequency used for this purpose is typically 50 or 60 Hz, depending on the country. People are exposed to LF fields around electrical and electronic devices, mainly at home and at work. Power lines are also a source of LF fields [6–9].

Radio frequency (RF EMF) is a term used to describe the part of the electromagnetic spectrum that encompasses the frequency range from 100 kHz to 300 GHz. Exposure to RF electromagnetic fields is usually measured in watts per square meter (W/m^2) or watts per kilogram (W/kg). After several decades of research on RF-EMF regarding numerous potential health effects, the only confirmed effect of RF EMF exposure significant for human health and safety is tissue heating. RF EMF fields can penetrate the body and cause vibrations of charged or polar molecules within. This results in friction, and thus heat. The body can tolerate a small increase in heat, similar to how excess body heat is dissipated during sports activities. This is because the human body has a strong ability to regulate internal temperature. However, above a certain level (known as the threshold), depending on the duration of exposure, exposure to radio frequency electromagnetic fields and the accompanying temperature increase can cause serious health effects, such as heat stroke and tissue damage [7–10]. Basic limits in the radio frequency range from 10 MHz to 10 GHz have been specified in the European Union Recommendation 1999/519/EC through the Specific Absorption Rate (SAR), expressed in W/kg. The SAR is a measure of the rate at which electromagnetic energy is absorbed and converted into heat in human body tissues, essentially indicating the power absorbed per unit mass of body tissue. When electromagnetic fields penetrate the human body, a portion of this field is absorbed and converted into heat within the tissues.

Recommendation 1999/519/EC allows for a SAR value 50 times lower than that which would cause an increase in human body temperature by 1 °C, and in the case of occupational exposure, 10 times lower [11]. On January 1, 2020, the Regulation of the Minister of Health of December 17, 2019, regarding the permissible levels of electromagnetic fields in the environment came into effect. This replaced the Regulation of the Minister of Environment of October 30, 2003, concerning permissible levels of electromagnetic fields in the environment and methods for maintaining these levels, specifically in terms of the threshold values for electromagnetic field impact [12–15]. With the introduction of the new Regulation, changes were made to the permissible parameters for fields with frequencies from 2 GHz to 300 GHz. The values were changed for the electric component from 7 V/m to 61 V/m (an increase of nearly nine times) and for power density from 0.1 W/m^2 to 10 W/m^2 (a 100-fold increase) [13, 14].

4 The Impact of Electromagnetic Fields on Humans

The impact of electromagnetic fields on the human body can cause the stimulation of excitable tissue due to the induction of electric currents in the body—this phenomenon is most significant at frequencies below several hundred kHz. Additionally, heating of tissues caused by the absorption of field energy, known as the thermal effect, is most significant at frequencies above 1 MHz. Furthermore, movement within an electromagnetic field can cause symptoms such as dizziness, nausea, and impaired visual-motor coordination, which cease upon moving away from the source. The effects of electromagnetic fields can lead to undesirable changes in body functioning. Scientific research indicates that potential effects of electromagnetic field exposure include disturbances to the nervous, cardiovascular, and immune systems, cancerous processes, and subjective ailments such as headaches, fatigue, and memory disturbances [15, 16].

Even 2 mG, or 0.2 μT, is sufficient to disrupt biological functions in humans. The maximum dose that a human can absorb is about 0.1 μT. People living near power lines, electricity generators, and similar disturbing devices report a high number of ailments. Research conducted in 1979 by physicist Ed Leeper showed that children exposed to just 0.3 μT were twice as likely to develop leukemia. This study was repeated in 1998 by the New York State Department of Health, confirming the previous findings [17]. Studies show that the protective threshold for the direct impact of alternating current electric fields should not exceed 5 V/m with grounding potential and 1.5 V/m without grounding potential. The emission of alternating current magnetic fields should not exceed 0.1 μT (1 mG). The radiation of radio frequency electromagnetic fields should not exceed 10 mW/m^2. These values are considered safe, but the protective threshold according to legal regulations on permissible levels of electromagnetic fields is much higher [18, 19]. A problem concerning the negative impact of electromagnetic fields on health is the fact that the effects of EMF exposure have a long latency period. The effects may only become apparent after 20–30 years. The development of brain cancer can take up to 40 years [16–18].

The impact of electromagnetic fields is not sensed by human senses. Direct perception of electromagnetic fields may occur when a person is in strong magnetic or electric fields of low frequencies, which can cause visual sensations. Moreover, moving within a magnetostatic field induces currents in the body, which can cause dizziness, nausea, and impaired visual-motor coordination. These effects cease after moving away from the magnetostatic field source and have an undetermined long-term impact on health, but they can significantly limit the ability to perform precise work. The impact of electromagnetic fields can cause undesirable changes in body functioning. Scientific research indicates that possible long-term effects of electromagnetic field exposure, especially in the case of strong exposure, include disturbances of the nervous, cardiovascular, and immune systems, cancerous processes, and subjective ailments such as headaches, fatigue, and memory disturbances [20–22].

5 Research Object and Testing Apparatus

The object of the study was an electric vehicle, the Dacia Spring (Fig. 1). The Dacia Spring was selected due to its relevance as a widely accessible and representative model in the growing market of electric vehicles. This choice aligns with the study's aim to investigate typical electromagnetic field emissions from electric vehicles commonly used in urban and suburban settings. The technical specifications of the model are presented below in Table 1.

Fig. 1. Dacia Spring electric (Author's own work).

Table 1. Technical specifications of the Dacia Spring [22].

Specification	Details
Model	Dacia Spring Crossover
Drive	Electric
Engine	Synchronous with permanent magnet
Engine Power	33 kW, 45 hp at 3000–8200 rpm
Torque	125 Nm at 500–2500 rpm
Engine Placement	Front, transverse
Transmission Type	Automatic, 1-speed
Drive Type	Front-wheel drive
Battery Type	Lithium-ion
Gross Battery Capacity	27 kWh
Charging Time (standard socket)	13 h
Charging Time (fast charging)	1 h
Energy Consumption	13.9 kWh/100 km
Range	230 km
Top Speed	125 km/h
Curb Weight	920 kg
Production Year	2021

Table 2. Technical specifications of the Dacia Spring (Source: [23]).

Feature	Details
Measured Field	Radio frequency electromagnetic field 10 MHz ~ 8 GHz
Measurement Method	3-axis digital measurement
Directional Characteristic	Isotropic, tri-axial
CAL Factor	Adjustable range 0.10~9.99 (default 1)
Measurements	CW Signal (f >50 MHz)
E (Electric Field)	38.0 mV/m; 11.0 V/m
H (Magnetic Field)	53.0 μA/m; 28.64 mA/m
S (Power Density)	0.1000 $\mu W/m^2$; 309.3 mW/m^2;
Dynamic Ranges	0.1000 $\mu W/cm^2$; 30.93 $\mu W/cm^2$
Accuracy	75 dBm
Isotropy Deviation	Absolute error ±0.1 dBm for 1 mV/m and 2.45 GHz
Measurement X, Y, Z	±1 dBm for 1 mV/m and 2.45 GHz

Measurements were conducted using the TM-196 high-frequency electromagnetic field meter. The TM-196 is a device operating in the range of 10 MHz to 8 GHz. Table 2 presents the technical parameters of the meters, and the measuring device is shown in Fig. 2.

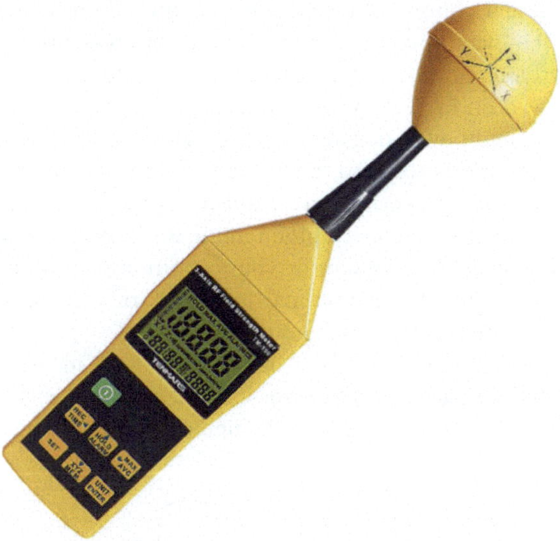

Fig. 2. TM-196 electromagnetic field meter (Source: [23]).

6 Electromagnetic Field Measurements in the Electric Car

Electromagnetic field measurements were conducted in the Dacia Spring electric vehicle to assess emissions under various driving conditions. The study included four distinct driving scenarios: urban traffic at approximately 30 km/h, higher-speed urban or suburban traffic at 50 km/h, moderate-speed highway traffic from 50 to 100 km/h, and high-speed highway traffic above 100 km/h up to 140 km/h. Each scenario involved 20 measurements to ensure statistical reliability and accuracy.

The measurements were taken during the afternoon and evening, maintaining a consistent ambient temperature of around 15 °C. Conditions were dry and windless, which minimized the influence of external environmental factors on the measurement results. The electromagnetic field meter was securely mounted inside the vehicle near the driver's seat to accurately capture the emissions experienced by the driver. To further ensure the accuracy of the data, all electronic devices within the vehicle, including the radio, were turned off during the measurement process.

The urban traffic scenario at 30 km/h represented typical city driving conditions with frequent stops and starts, traffic lights, and pedestrian crossings. The higher-speed urban or suburban traffic scenario at 50 km/h involved driving in less congested areas, such as residential neighborhoods or light traffic zones in the suburbs. For the moderate-speed highway traffic scenario, the vehicle-maintained speeds between 50 and 100 km/h on highways with moderate traffic flow. The high-speed highway traffic scenario assessed electromagnetic emissions at speeds above 100 km/h, typical of highways or expressways.

These structured and controlled measurements aimed to provide a comprehensive understanding of the electromagnetic emissions in different driving conditions. The flowchart of the research process is presented in Figs. 3 and 4.

The results showed on Figs. 5, 6 and 7 indicated that the emissions of the electric field, magnetic field, and power density increased with the load on the electric motor. Specifically, at maximum motor load, the electric field values reached approximately 4 V/m (Fig. 5). This finding is critical as it illustrates the direct relationship between motor load and field emission intensity, highlighting the importance of monitoring electromagnetic emissions during different driving conditions. These measurements provide valuable insights into the operational behaviors of electric vehicles and their electromagnetic emissions under varying conditions.

In addition to driving conditions, electromagnetic field measurements were also conducted during the vehicle charging process. These measurements were performed at the designated charging station located in the underground parking lot of Silesia City

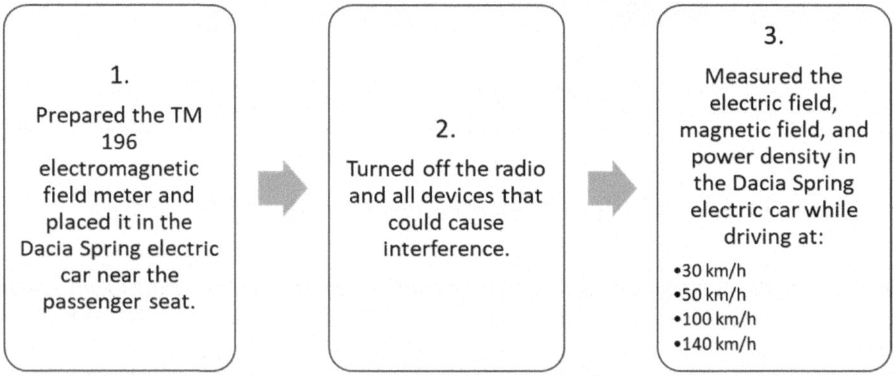

Fig. 3. Flowchart of the elements of research while driving the electric vehicle (Author's own work).

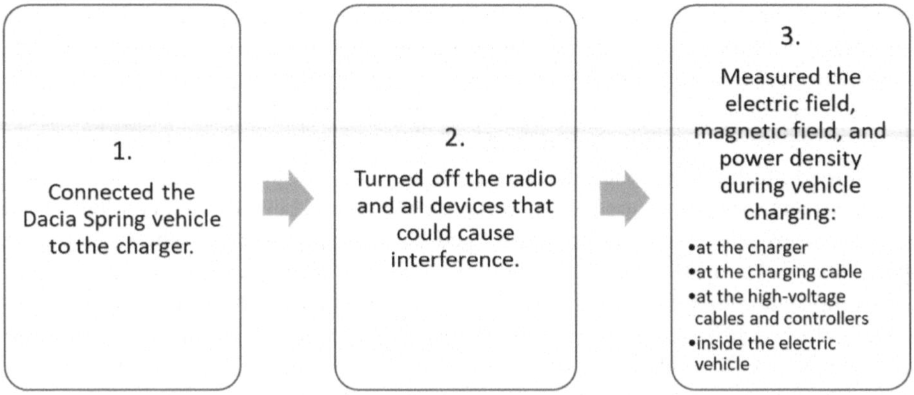

Fig. 4. Flowchart of the elements of research while charging the electric vehicle (Author's own work).

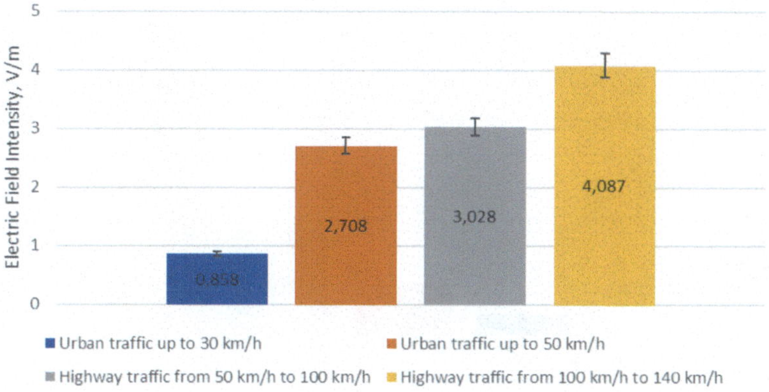

Fig. 5. Electric field intensity inside the vehicle—electric drive (Author's own work).

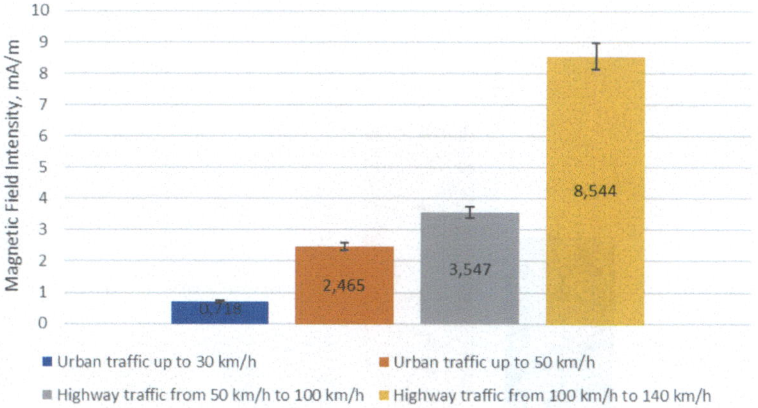

Fig. 6. Magnetic field intensity inside the vehicle—electric drive (Author's own work).

Center. Each measurement session at Silesia City Center involved a systematic and comprehensive approach, comprising a total of 20 individual measurements. This was done to ensure the reliability and repeatability of the results, thereby providing a robust dataset for analysis. The measurements were conducted during the afternoon and evening hours, chosen to reflect typical usage times and to maintain consistency across all sessions.

The ambient temperature during these sessions was approximately 15 °C, a condition that was constant to ensure that temperature variations did not influence the measurement outcomes. The weather conditions were also consistently dry and windless, eliminating any potential external environmental factors that could affect the accuracy of the measurements. To further enhance the precision of the data collected, meticulous steps were taken to minimize measurement error. All electronic devices within the vehicle, including the radio and any other non-essential equipment, were turned off during the measurement process. This precautionary measure was crucial to ensure that the

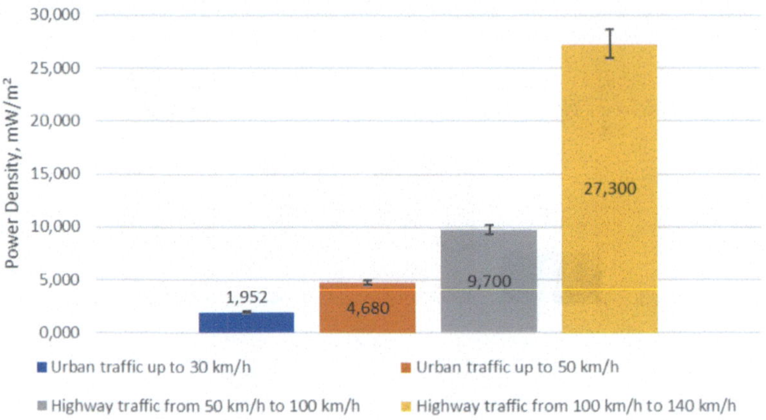

Fig. 7. Power density inside the vehicle—electric drive (Author's own work).

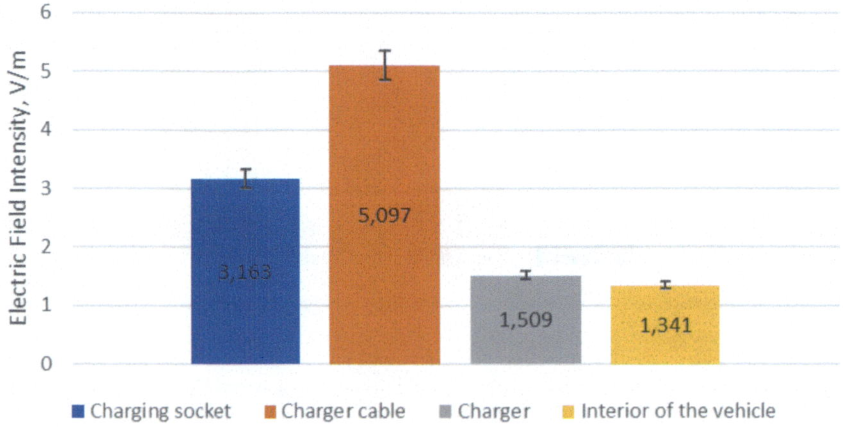

Fig. 8. Electric field intensity during electric vehicle charging (Author's own work).

electromagnetic fields measured were solely due to the vehicle's charging process and not influenced by other electronic interferences.

The electromagnetic field meter was positioned strategically within the vehicle, ensuring optimal detection of emissions from the charging system. The careful placement of the meter, coupled with the controlled environment of the underground parking lot, provided a consistent and reliable setting for capturing the electromagnetic emissions associated with the vehicle charging process. The results of the measurements are shown in Figs. 8, 9 and 10.

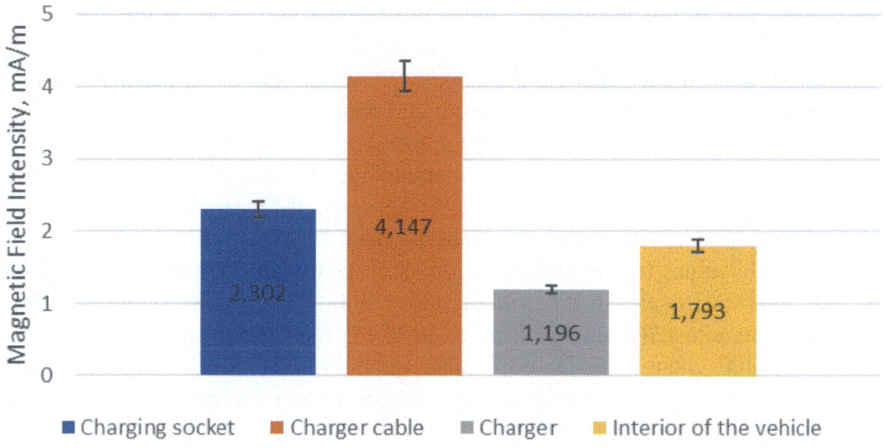

Fig. 9. Magnetic field intensity during electric vehicle (Author's own work).

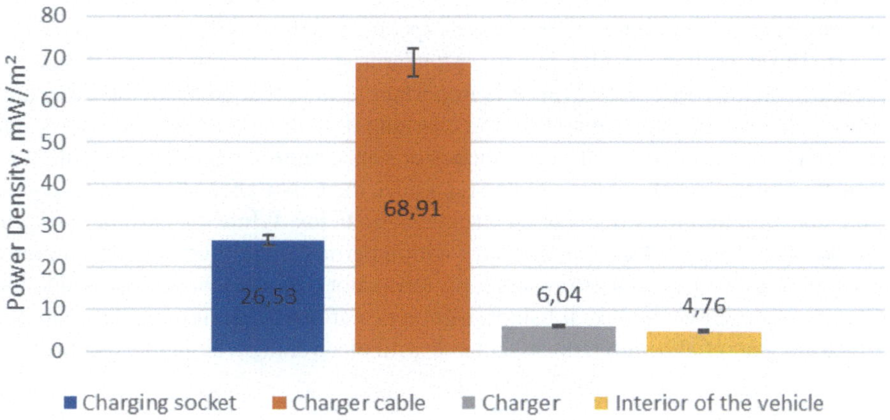

Fig. 10. Power density during electric vehicle charging (Author's own work).

7 Summary

Measurements of the electromagnetic field in the Dacia Spring electric vehicle revealed that the emissions of the electric field, magnetic field, and power density increase in correlation with the load on the electric motor. Detailed observations showed that the electric field values at maximum motor load reached approximately 4 V/m. This finding is critical as it illustrates the direct relationship between motor load and field emission intensity. Additionally, during the charging process, the maximum electric field recorded was around 5 V/m Additionally, the magnetic field and power density increased proportionally with motor load, as shown in Figs. 5–10, demonstrating significant emission levels during both driving and charging conditions. These results highlight the importance of continuous monitoring to ensure safety and regulatory compliance. On January 1, 2020, the Regulation of the Minister of Health of December 17, 2019, regarding permissible levels of electromagnetic fields in the environment came into effect. This regulation significantly amended the previous guidelines set by the Regulation of the Minister of Environment dated October 30, 2003, which addressed permissible levels of electromagnetic fields in the environment and methods for maintaining these levels. The amendments particularly focused on the threshold values for the impact of electromagnetic fields, reflecting the evolving understanding and technological advancements in this field. With the introduction of the new regulation, the permissible parameters for electromagnetic fields with frequencies ranging from 2 GHz to 300 GHz were revised. The permissible values for the electric component were notably increased from 7 V/m to 61 V/m, marking an almost ninefold increase. Similarly, the permissible power density values were adjusted from 0.1 W/m^2 to 10 W/m^2, representing a 100-fold increase. These substantial changes underscore the regulatory shift towards accommodating higher field intensities, potentially due to advancements in technology and a deeper understanding of electromagnetic field effects [13, 14].

Studies conducted on the electromagnetic field generated by the electric motor and high-voltage cables within the electric vehicle, as well as those emitted by the charger, indicated that these emissions did not exceed the values deemed permissible for the protection of health and safety of workers as specified in the Act Dz. U. 2019 poz. 2448 [12, 13]. This compliance with regulatory standards signifies that the use of the electric vehicle is generally safe. However, it is important to note that being in the immediate vicinity of the vehicle, particularly during charging or high motor load operations, may cause temporary ailments such as headaches or concentration disorders. Despite these optimistic research results, it is crucial to recognize the importance of awareness regarding the potential harmful effects of electromagnetic fields. While the current emissions from the Dacia Spring electric vehicle are within permissible limits, ongoing monitoring and research are necessary to ensure long-term safety. Implementing strict guidelines for exposure durations and establishing safe zones around charging stations can help mitigate any potential health risks associated with electromagnetic field exposure.

References

1. Raczkowski, B.: BHP w praktyce, 14th edn. ODDK Press, Gdańsk (2012)
2. Koradecka, D.: Bezpieczeństwo i higiena pracy. CIOP Press, Warszawa (2008)
3. Augustyńska, D., Pośniak, M.: Czynniki szkodliwe w środowisku pracy wartości dopuszczalne. CIOP Press, Warszawa (2003)
4. Aniołczyk, H., Bartkiewicz, A., Gadzicka, E., Gryz, K., et al.: Pole elektromagnetyczne źródła – oddziaływanie – ochrona. Instytut medycyny pracy im. prof. J. Nofera, Łódź (2000)
5. National Advisory Committee for Aeronautics: https://www.nasa.gov/wp-content/uploads/2009/07/319904main_The_Electromagnetic_Spectrum.pdf
6. Przystupa, P., Pohrebennyk, V., Mitryasova, O., Kochan, O.: The influence of electromagnetic field on human health. Conference: 2019 Applications of Electromagnetics in Modern Techniques and Medicine (2019)
7. Mercola, J.: Pole Elektromagnetyczne EMF. Vital Press, Białystok (2020)
8. Wertheimer, N., Leeper, E.: Electrical wiring configurations and childhood cancer. Am. J. Epidemiol. **109**, 273–284 (1979)

9. Wartenberg, D.: Residential magnetic fields and childhood leukemia: a meta-analysis. Am. J. Public Health. **12**, 1787–1794 (1998)
10. Frey, A., Feld, S., Frey, B.: Neural function and behavior: defining the relationship. Ann. NY Acad. Sci. **247**, 433–439 (1975)
11. Parathath, S., Parathath, S., Tsirka, S.: Nitric oxide mediates neurodegeneration and breakdown of the blood-brain barrier in tPA-dependent excitotoxic injury in mice. J. Cell Sci. **119**, 339–349 (2006)
12. Tang, J., Zhang, Y., Yang, L., Chen, Q., Tan, L., Zuo, S., Feng, H., Chen, Z., Zhu, G.: Exposure to 900 MHz electromagnetic fields activates the mkp-1/ERK pathway and causes blood-brain barrier damage and cognitive impairment in rats. Brain Res. **1601**, 92–101 (2015)
13. Regulation of the Minister of Health of the Republic of Poland of 17 December 2019 on permissible levels of electromagnetic fields in the environment. https://isap.sejm.gov.pl/isap.nsf/DocDetails.xsp?id=WDU20190002448
14. Regulation of the Minister of the Environment of the Republic of Poland of 30 October 2003 on the permissible levels of electromagnetic fields in the environment and ways of checking whether these levels are met. https://isap.sejm.gov.pl/isap.nsf/DocDetails.xsp?id=WDU20031921883
15. European Union Community Research and Development Information Service. https://cordis.europa.eu/article/id/156586-electric-cars-no-risk-for-human-health/pl
16. Zhao, L., Liu, X., Wang, C., Yan, K., Lin, X., Li, S., Bao, H., Liu, X.: Magnetic fields exposure and childhood leukemia risk: a meta-analysis based on 11,699 cases and 13,194 controls. Leuk. Res. **38**, 269–274 (2014)
17. Gumiela, J., Stanik, L., Sztafrowski, D.: Pomiarowa identyfikacja emisji pola elektromagnetycznego przez samochód elektryczny. Przegląd Elektrotechniczny. **95**, 128–131 (2019)
18. Verheijen, E., Jabben, J.: Effect of electric cars on traffic noise and safety. Report of National Institute for Public Health and the Environment (2010)
19. Gryz, K., Karpowicz, J., Zradziński, P., Tokarski, T., Kapica, L.: Ekspozycja na pole elektromagnetyczne podczas użytkowania pojazdów samochodowych z napędem elektrycznym lub hybrydowym, vol. 12, pp. 18–21. Centralny Instytut Ochrony Pracy – Państwowy Instytut Badawczy Press (2020)
20. 1999/519/EC: Council Recommendation of 12 July 1999 on the limitation of exposure of the general public to electromagnetic fields (0 Hz to 300 GHz). https://www.prawo.pl/akty/dz-u-ue-l-1999-199-59,68183446.html
21. Communications Institute, National Research Institute. https://www.gov.pl/web/5g/pomiary-pem-w-srodowisku-ogolnym%2D%2Dzagadnienia-wybrane
22. Operating instructions for the Dacia Spring. https://www.dacia.pl/obsluga-techniczna.html (06.12.2023 r.)
23. Operating instructions for the TM196 MERASERW power and field strength meter. https://www.meraserw5.pl/images/Tenmars/TM196_DTR.pdf

LCA for Off-Road Vehicles Intended for Persons with Disabilities

Marcin Staniek[1] (✉) and Ireneusz Celiński[1]

[1] Silesian University of Technology, Faculty Transport and Aviation Engineering, Gliwice, Poland
marcin.staniek@polsl.pl; ireneusz.celinski@polsl.pl

Abstract. As part of a competition organised by Narodowe Centrum Badań i Rozwoju [National Centre for Research and Development], entitled *Rzeczy są dla ludzi* [*Things are for People*], the Faculty of Transport and Aviation Engineering of the Silesian University of Technology received funding for the following project: *MOUNTAINS WITHOUT BARRIERS—Integrated platform for the planning, organisation and oversight of as well as support for the access to mountain tourism services for people with physical difficulties in functioning, and a special-purpose off-road vehicle intended to support the provision of tourism services.* Special-purpose off-road vehicles for people with special needs are becoming increasingly common around the world. They are two-track equivalents of mountain bicycles, designed for persons with special needs. These vehicles feature highly specialised factory-installed drivetrain tailored to the user's individual needs, which is frequently electrically assisted. These vehicles are often equipped with additional special-purpose components. As any other vehicles exposed to major kinematic loads, these are also subject to increased wear and tear, for instance of mechanical parts and electric batteries. This article represents an attempt to address the overall body of problems pertaining to the life cycle assessment of such vehicles.

Keywords: LCA · Persons with disabilities · MTB · Electric battery

1 Introduction

Companies manufacturing high-end off-road vehicles for people with special needs (OVfPwSN) have been present in Poland for more than a dozen years (the first one was founded in 2008) [1, 2]. The situation is similar in many other countries, including USA, UK, Canada, etc. [3–5]. Polish companies are major and globally present manufacturers of such vehicles, their fair share being exported. An increasing number of providers of such solutions, intended for people with special needs, are establishing operations in further countries, including Poland. They have noticed the existing need in this sphere as well as the significant, yet not fully recognised, potential of the market, which is growing rapidly and can be considered potentially very capacious (tens of millions of users worldwide). According to the Government's Representative for Persons

G. Sierpiński et al. (Eds.): TSTP 2024, LNNS 1257, pp. 107–128, 2024.
https://doi.org/10.1007/978-3-031-81777-9_8

with Disabilities, the maximum size of the special needs group in Poland alone is estimated at 4.9–7.7 million persons. However, within this community, the population of people who may be attributed the direct demand for such devices is estimated at ca. 2 million (as per the 2021 National Census). The potential buyers of OVfPwSN represent a fraction of that group that cannot be precisely defined. Nevertheless, even assuming that, over the next 10 years, merely every 20th person within this group purchases such a vehicle, the value of this market can be estimated at approx. PLN 2.5 billion. This value is nearing that of an annual revenue of a company operating on a country-wide scale in the sector of home appliances (as of 2022).

At the same time, it is a difficult market, and for many reasons, including the need for customisation of vehicles, the relatively high cost of purchase, the use of high-end processing technologies in production, an underdeveloped rental market, and a virtually non-existent after-market. Prices of many new vehicles begin at PLN 25,000 (EUR 6000, as at June 2024), which makes them similar to those of used motor vehicles of low and medium engine capacity [1]. A well-equipped and functionally advanced vehicle of this type can cost several times more. From the point of view of the owner's place of residence, what is required each time for the vehicle to be operated is its transport to the specific place where mountain areas are to be explored. This entails using a motor vehicle with large enough storage space. Selected OVfPwSN models can be folded for transport purposes in such a manner that they even fit into some versions of estate and hatchback cars. The public transport system, given its current organisation and the functionality of means of transport, does not make it any easier to move such vehicles over long distances. The range of the off-road vehicles currently in manufacture is really large, spanning both recreational and high-performance cycles, as well as those designed for elderly people and those suited for children. What individual hardware solutions also take into account is the specificity of a given user's needs in terms of the drivetrain intended for such means of transport [1–5]. With the foregoing in mind, and in light of the various challenges observed in this market, life cycle assessment (LCA) based analyses can be perceived as an important problem, which is precisely why they have attracted the authors' attention.

Life cycle assessment is a methodology aimed at estimating the environmental hazards arising from the very existence as well as the use or operation of a specific product. LCA is a process in which (for potential factors) one assesses the effects a given product exerts on the environment. However, by no means are assessments of this kind obvious. As the name itself implies, LCA pertains to the entire *life* of a specific product. In this context, authors of various publications on the subject often use the term *from cradle to grave* [6–11]. The problem is that defining where a product's life starts is purely conventional. Likewise, certain consequences of impacts persist much longer than the product itself lasts (extending also past its formal and legal disposal). For instance, it takes up to 1000 years for plastic and up to 4000 years for glass to decompose. LCA can pertain to a product or its specific feature. The process is based on the identification and quantitative assessment of a product's burdens in terms of its effects on the environment [6–11]. What follows is an evaluation of the potential environmental impacts of the burdens previously identified. The deliverable of the foregoing two stages may and should be the identification of measures aimed at reducing the identified environmental risks. The identification of the environmental burdens of a product comprises such product-related aspects as the materials and energy consumed, or the emissions released into the environment, etc. [6–11]. Except perhaps for the first group of factors, it is often difficult to identify the exact spatiotemporal parameters of these problems. The energies consumed and the emissions released may relate to areas considered on a global scale (which is not always taken into account). Properly implemented, the LCA procedure enables a legitimate assessment of the environmental impact of a given product. The method in question has been described in the relevant ISO standards [12, 13]. LCA comprises four stages: goal and scope definition, inventory analysis, i.e. defining the set of inputs and outputs, impact assessment, and finally the assessment proper [12, 13]. Nevertheless, practice shows that this can be approached in multiple ways.

Managing what is commonly referred to as environmental relations of businesses is covered by the ISO 14000 family of standards [12, 13]. These are international standards for environmental management, implying tools for regulating environmental aspects [12, 13]. The following are further standards of the ISO 14000 series. ISO 14040:2006 describes the principles and framework (structure) of life cycle assessment. The document provides the definition of the purpose and scope of LCA, and further defines the life cycle inventory (LCI), the life cycle impact assessment (LCIA), the cycle interpretation stage, the LCA reporting and critical review, the limitations of LCA, the relationship between individual LCA stages, as well as the conditions for the use of value choices and optional elements [12]. Importantly, the standard neither describes in detail the LCA technique itself, nor specifies any methodology intended for individual stages. It should be emphasised that, while defining the purpose and scope, one should consider the intended application of the LCA results [12]. However, the application itself is beyond the scope of the standard in question [12]. ISO 14044:2006 defines the relevant requirements and provides guidelines for life cycle assessment. In this respect, it takes into account the definition of the LCA purpose and scope, the life cycle inventory (LCI) stage, the life cycle impact assessment (LCIA) stage, the cycle interpretation stage, the LCA reporting and critical review, the limitations of LCA, the relationship between individual LCA stages, as well as the conditions for the use of value choices and optional elements [13]. The ISO 14050:2020 standard concerns the vocabulary used in relevant documentation. It defines the terms used in the field of environmental management systems. Another standard pertains specifically to the environmental management systems (EMS) themselves (ISO 14001/14004), covering the design and implementation of the EMS. The aspect of continuous improvement of businesses in terms of environmental protection is considered particularly important in this respect. The standard describes various problems of resource use and waste management. ISO 14004 provides general guidelines for the implementation of an environmental management system. These guidelines are intended for an organisation which manages its environmental responsibilities in a systematic manner. The ISO 14031:2021 standard concerns environmental performance assessment and provides guidance on the design and application of environmental performance evaluation (EPE), regardless of the company type, size and complexity. The document does not specify the levels of environmental performance. ISO 14015:2022 addresses the assessment of environmental due diligence. ISO 14049:2012 provides examples of practices for the performance of the life cycle inventory (LCI) analysis to meet specific requirements of ISO

14044:2006. ISO 14025:2006 lays down principles and defines procedures for developing environmental declaration programmes. ISO 14025:2006 establishes principles for the use of environmental information. The environmental declarations described in ISO 14025:2006 are intended for use in business-to-business communication. The ISO 14064 standard provides organisation-level guidance for quantifying and reporting greenhouse gas emissions and removals. It is there that specific principles and requirements concerning the quantification and reporting of GHG emissions and removals have been provided. What matters in this respect is the system known as EMAS (Eco-management and Audit Scheme), based on Regulation (EC) no. 1221/09 on the voluntary participation by organisations in the eco-management scheme.

2 LCA for E-Bikes and E-Scooters—A Literature Review

The review of the literature on the subject begins with electric bicycles. These vehicles represent a major group of means of transport in Southeast Asia, the USA (mainly in metropolitan areas), and certain European countries [15, 16]. Contrary to appearances, they are not zero-emission vehicles. Using the production and operation of these vehicles in China as an example, selected environmental impacts of electric bicycles have been quantified [16]. And since they are still a rarity among bicycles, an important contribution of that study to the state of the art is the comparison of these vehicles with those of analogous impacts, i.e. buses, motorcycles, and cars. The authors of the said article have proved that, in statistical terms, electric bicycles generate significantly less pollution per kilometre than motorcycles and cars. However, importantly, their emissions are similar to those of buses. The use of lead-acid batteries causes electric bicycles to release dangerous lead, which represents the main issue in this case [15]. It is noteworthy that not everywhere is one bicycle considered equivalent to one passenger vehicle in traffic. In Poland, this ratio ranges at 1.2–1.6, depending on the urban area. Adequate traffic management should reduce the unfavourable passenger car to bicycle ratio (tandems still account for a trace market share).

Some further studies have added to the discourse on the subject of interest addressed in this article [15]. They provide an analysis of 638 surveys, completed mainly in Europe and North America. The survey results indicate that the following factors are considered important by potential users of electric bicycles: safety, comfort, climbing assistant, brake lights, and brake indicators. Those surveyed generally declared an intent to own a smarter, safer, and more adaptive vehicle [15], which is precisely what can be said about the off-road vehicle for people with special needs (OVfPwSN) developed under the authors' project.

Another article addresses an LCA of a micro-mobility vehicle known as Ghisallo, being basically a tricycle typically intended for urban applications. The design of the steering system, suspension and frame of this tricycle makes it actually possible to use public means of transport, such as trains, buses, underground railway, etc. This makes the solution similar in conceptual terms to the idea of carrying OVfPwSN-type vehicles by public transport. With regard to the said vehicle, i.e. Ghisallo, the environmental impact assessment method proposed by the International Transport Forum was used.

The tricycle was upgraded in terms of the materials used and the energy source that served for battery charging. The operational results thus achieved were similar to the values obtained by manufacturers of other micro-mobility vehicles. The energy consumption and greenhouse gas emissions (grammes of CO_2 equivalent) per vehicle kilometre came to 0.36 MJ/v-km and 29 g of CO_2eq/v-km, respectively [17]. The authors of that study concluded that the majority of greenhouse gas emissions came from the tricycle production (42% of total emissions). Air transport from the place of manufacture to sales locations increases the environmental impact by further 10%. The authors have proposed specific measures aimed at reducing the life-cycle impact of this micro-mobility vehicle on energy consumption and GHG emissions [17]. They call for increased precision of the LCA inventory stage through contacts with manufacturers, mainly on account of the need for determining the actual share of recycled materials in the production [17]. Another aspect in question is the transport of the finished product. Defined in the article as a recommendation of somewhat secondary importance, still very relevant, is the use of cycle paths at the level of 0% or 90%, which is conducive to reducing environmental impact. The authors believe that this factor requires verifications in further studies [17]. This is more of a topic for BCA studies for the transport network infrastructure. They have noted that shared micro-mobility has a negative impact on performance, mainly by reducing the service life of vehicles, as well as on the provision of the required operation-related services (additional, compared to those functioning outside the shared mode of this means of transport) [17].

In the sphere of micro-mobility, including mainly shared mobility, electric scooters are used besides electric bicycles. Another article addresses various problems of electric scooters [18], and it discusses studies performed in Germany. They included a survey conducted among users of this modern means of transport as well as interviews with experts. The latter were asked whether they considered electric scooters an environmentally friendly means of transport. In physical terms, the environmental impact of these vehicles was determined by way of an LCA. The impact of the scooters calculated under the study came to 165 g of CO_2 equivalent/km, and it was mainly attributable to the materials used as well as the production, which altogether accounted for 73% of the impact [18]. The results presented in that article are similar to those obtained under the survey conducted in China [16]. Therefore, one can conclude that it is mainly the production of these means of transport that poses the problem, followed by their operation and potential transport to the place of use. Electric scooters with replaceable batteries reduce the global warming potential by 12%. Assuming that the service life of electric scooters is extended to at least 15 months, the CO_2 equivalent/km of 46 g will be achieved [18]. Therefore, what also matters for purposes of the LCA analysis of OVfPwSN is that, during the operation of these vehicles, long-distance transport is often involved (which is due to the high specialisation of production and the number of manufacturers being still relatively small). There is a company which specialises in the production of vehicles for anglers and hunters. Such a vehicle sometimes travels thousands of kilometres before it reaches the recipient. Moreover, the number of components used in production is greater than in the case of single-track vehicles. For the sake of depiction, one can argue that a conventional bicycle, and even a mountain bike, represents a—more or less, but still—flat design, while that of the OVfPwSN is developed in a 3D space. What also matters is the aspect of batteries and accessories being replaceable, which is generally the option preferred by bicycle users.

There are typically two kinds of bicycle sharing systems in common use: those based on docking stations and dockless (i.e. free floating, without set location). Their carbon emission profiles differ in terms of production, operation and maintenance, as well as disposal [19]. Against such a context, the authors of the referenced article conducted an LCA based study of shared bicycles for Capital Bikes in Washington, D.C. (USA) [19]. In the case analysed, it was the installation of platforms and docking stations that represented the main source of carbon emissions for the bicycles. On top of that, also fleet management and maintenance constituted major emission sources for these vehicles. The authors of the study in question also argue that manufacturing operations generate five times more carbon dioxide emission than disposal related activities. The results obtained under the study were compared with a total normal impact. The Capital Bikes case study allowed the authors to conclude that in order for each bicycle to ensure greenhouse gas emission optimisation it should be used for at least 591 days. Therefore, one should aim at extending the service life of bicycles, which is a very complex problem in itself [19]. Considering the body of problems at hand, a new theme appears to be noteworthy, namely the installation of the shared bicycle infrastructure (and more broadly, not just docking stations, but also other pieces of the infrastructure, such as replacement batteries, as for the OVfPwSN described herein).

Another LCA based case study concerning means of micro-mobility is one which covers Ampler Bikes [20]. It is a company that manufactures electric bicycles, focusing on their optimised design and durability over the lifetime. The company conducted an LCA concluded with a published report [20]. As part of the analysis in question, the carbon footprint of the Stellar bicycle model was calculated with the support of the GreenDelta computational software [21]. The results of such studies are typically used to identify the most important areas of business operations and to set emission reduction targets for 2025 (the company's goals) [21]. The bicycle manufacturer and the author of the report have suggested that the entire bicycle and mobility industry needs to be changed in terms of environmental emissions. The report and the analysis are based on a conventional set of problem fields: materials, design, production, assembly, operation, and disposal [20]. The GHG emission results have been provided in a breakdown into the following categories: production of components, transport of components, assembly, picking and packing, product transport, operation, and disposal [20]. The following emissions have been stated in the report: 440 (kg of CO_2 eq.) for the operation phase (54%), 223 for production of components (27%), 101 for transport of components (12%), 29 for assembly (4%), 10 for the packaging process, and 6 for each: transport to the destination point and disposal [20]. Importantly, the manufacturer has presented these figures broken down by subassemblies, and so the following emissions (kg of CO_2 eq.) have been reported, respectively: 33—frame, 25—charger, 16—wheel rims, 14—suspension forks, 13—battery, 13—crankset mechanism [20].

Another carbon footprint study was conducted for a case of in-campus transport. It covered buses and a bike sharing system used to complete identical origin-to-destination transfers. Under that research, LCA was used with regard to a case study of four campuses: College Ave, Cook/Douglass, Busch, and Livingston at Rutgers University (New Brunswick, New Jersey) [22]. The relevant life cycle analyses were broken down into the following stages: raw material acquisition, production, transport, operation and maintenance, and end-of-life. The analysis uncertainty factors, i.e. fossil fuel type, number of bicycles, and frequency of bus use, were identified as the sensitivity variables of the analysis performed. The normalisation method was employed under two impact categories. The authors have concluded that the majority of CO_2 emissions and energy consumption are related to the raw material stage (upstream extraction and production), the bike sharing system, and the operation of the campus bus system. The CO_2 emissions and the energy consumption by the campus buses were, respectively, 46 and 13 times that of the bike sharing system proposed by the authors [22], who have further argued that biodiesel fuel can significantly reduce the CO_2 emissions and energy consumption within the existing campus bus system, and that an increased number of bicycles increases the CO_2 emission from the bike sharing system, since a larger number of bus passengers affects both systems. According to the authors, a hybrid system, where 70% of passenger-kilometres would be covered by campus buses and the remaining 30% by a bike sharing system, is the most environmentally friendly setup [22]. This is actually one of the very few studies conducted in this area which address the problem of modal split optimisation. The author of yet another study points out that it is the materials used in the construction of a given single-track vehicle that enable understanding of its environmental impact. That publication contains a very detailed description of the technique and technology employed in the production of a single-track vehicle from aluminium and other raw materials, highlighting the role of the raw materials consumed in these processes and the logistic chains used to move them. It reads that the two most common bicycle repair activities are tyre and chain replacement. What the foregoing data imply is that both involve cumbersome materials, mainly due to the rubber (tyre) and the large amount of waste from chain production. Consequently, this results in consuming more synthetic rubber, natural rubber, and steel in order to maintain continuity of bicycle operation [23, 24]. The author of the article continues by claiming that the life cycle of an aluminium bicycle requires materials in large quantities, adequate transportation (logistic chains for components), proper maintenance, limited to key components, finally being subject to standard disposal (recycling) [24]. In conclusion, the author highlights the fact that, for purposes of the transport of the products used in the production of an aluminium bicycle, the amount of fossil fuels consumed is in fact very high [24]. Nevertheless, she repeatedly stresses that aluminium, which is common to the production of this means of transport, is easily moulded and recycled [24]. What should be noted in this respect is that there are many different aluminium alloys, especially those used in highly specialised vehicles (authors' note). This proves to be an important matter in the context of the problems at hand, since not only aluminium components are used in the production of OVfPwSN, but also metals such as titanium, offering resistance to corrosion and high strength @ low weight.

Another article analysed against the subject in question concerns the Specialized Bicycle Components brand [25]. An environmental study was conducted in this context. Also, a report on the quantitative analysis of the sustainability of the processes used in the production of the brand's bicycles was compiled. An LCA was carried out to quantify the impact of the various components of the said bicycles [25]. The Outdoor Industry Association's (OIA) equipment index was applied in the assessment. The OIA Equipment Index was intended as a tool enabling businesses to assess their sustainability—from the specific materials they use to overarching brand policies. The Equipment Index differs from LCA by assuming a broader and more qualitative approach that attempts to measure a company's progress toward sustainability. The LCA results have shown that carbon fibre composite requires consumption of significant amounts of water, and by contrast, the production of aluminium bicycles consumes large amounts of energy, while the production of bicycle chains generates considerable amounts of recyclable waste. As a side note, there have been attempts to produce bicycles from various atypical materials like plastic, bamboo, flax, recycled materials, or other fibres. Their outcomes, aside from the frequently questionable performance, do not show significant emission reductions. The index results favour alu as a bicycle frame material, but do not comprehensively indicate any specific impact typical of each product [25].

Another study raises an entirely different, yet important issue, which happens to be related to the OVfPwSN production sector. One of the recurring themes of the project addressed in this article is—as evidenced—the bicycle industry's concerns about privacy associated with the data collection by the production team. This is difficult to avoid in the industry on account of the need for customisation of many of the vehicles produced. The article highlights the importance of supply chains for production [25].

The author of yet another paper points out that a family or cargo electric bicycle has the potential to reduce emissions and lower the environmental footprint [26]. Accordingly, it presents an LCA study based on the ReCiPe impact assessment method, where calculations were performed using the SimaPro LCA software [27]. According to this approach, a long list of LCA inventories is converted into a limited number of indicator results. Consequently, it has been found that bicycle production has a significant effect on the environment (contrary to the general public's opinion on the subject). The author of the article has established the carbon footprint of a bicycle at 195.2 kg of CO_2 eq. This result is affected by the aluminium consumed in the manufacturing process. The steel used in bicycle parts has been identified as a factor relevant to the creation of toxicological burdens [26]. The author suggests that bicycle manufacturers should focus on environmentally friendly recycling of aluminium and batteries [26].

The bicycle is a massively used means of transport in Asia, especially in Southeast Asia. In this context, a framework for designing environmentally friendly products has been developed in Bangladesh [28]. The study in question addressed the ISO 14044 guidelines and used life cycle inventory data extracted from an article previously released [29]. The study results imply that it is production that exerts the greatest environmental impact [29]. Therefore, nearly every research completed in the subject matter area confirms the role of production in affecting the environment.

The goal of yet another study, also concerning the problem at hand, was to quantify the life cycle emissions of a conventional bicycle. To this end, an LCA was conducted using two different technological approaches: the OpenLCA product system modelling suite and the sustainable design module of the popular SolidWorks program [30–32]. The aforementioned software was used to analyse the parameters assumed to affect the LCA inventory stage. By that means, the relevant environmental burdens of the processes associated with the selected vehicle were determined [30]. The authors have demonstrated the differences in the use of the various IT tools. The result thus obtained for the SolidWorks software is 12.69% higher than that of OpenLCA. What the study further implies is that the factors dominant in terms of environmental impact are material extraction, the types of supplies for production, and the methods chosen for the processing of decommissioned bicycles [30]. The observation concerning the differences in the array of software tools used for LCA purposes is relevant on account of the fact that there is a number of categories related to the elements of a bicycle LCA (broken down by both components and production activities) that fall within the range threshold nearing ±13%.

The SolidWorks software is relatively often used for the sustainability analysis and life cycle assessment of bicycles [32, 33]. The analysis helps in determining which bicycle components have the greatest environmental impact. The SolidWorks software makes it possible to reduce the environmental impact of a component identified as problematic by improving the vehicle design [33]. The material used to build the entire assembly is evaluated in terms of performance, reliability, and environmental pollution by taking carbon footprint, energy consumption, air acidification, and eutrophication into account [33]. Although it represents an interesting research material, the use of wooden elements in OVfPwSN seems problematic on account of the dynamic loads observed [33].

The OVfPwSN type vehicles, perceived as environmentally friendly means of transport, can function at the same time as public shared cycles, yet to a limited extent. This limitation is due to the need for vehicle adaptation to the specific user needs. If such a cycle becomes customised to serve specific user categories, then by all means it can be rendered publicly available for hire (another study has been cited, showing that this changes its environmental impact significantly and negatively, although this depends on the vehicle's use period; see below). On account of the equipment and accessories of this vehicle, leaving it at docking stations is rather problematic (potential theft of components, damage, weather proofing that entails further costs). OVfPwSN type vehicles are much more advanced than regular cycles.

In another study, public bicycles were subject to an LCA analysis based on carbon emission calculations [34]. The authors have shown that, where the average daily turnover rate of public bicycles comes to 1.87, the average daily distance covered is 2.15 km, and the damage rate increases by 2.5% every month, then it takes about 7 months for each public bicycle to strike the carbon balance [34]. According to the authors of that study, once the carbon balance has been attained, further use of the public bicycles results in a net reduction of carbon dioxide emission [34]. Interestingly, the carbon dioxide emission exceeds the emission reduction again after ca. 29 months of operation [34]. Therefore, this impact is variable over the vehicle lifetime, and not monotonic. The case study discussed in the paper in question consisted in analysing the carbon dioxide balance of 186 bike sharing stations available in the city of Wenling, in the Chinese province of Zhejiang [34]. The authors have also proposed some policy recommendations intended to help in achieving actual emission reductions [34].

Some of the articles on the problem at hand concern the LCA analysis based on specific energy sources. One of such studies, based on the environmental life cycle assessment model aimed at evaluating electric bicycles powered by lithium-ion batteries against petrol engine cars and battery electric vehicles (BEV), was conducted in the UK [35]. What the study took into consideration was the use of this means of micro-mobility underpinned by a specific motivation, namely commuting. The analysis provided in the publication covered the vehicle life cycle and the fuel life cycle. The model also comprised emissions from tyre and brake wear as well as their impact on noise [35]. The results thus obtained imply that BEVs and electric bicycles reduce CO_2 emissions, and that this reduction can increase along with a change in the proportion of electricity consumption and adequate recycling of batteries. What matters from the point of view of this study is that the advantages of electric bicycles are not proportional to the vehicle weight (most OVfPwSN are much heavier). Non-combustion emissions exert a significant impact on the human respiratory system and the toxicity thus caused, and so does the noise reduction attributable to the use of electric vehicles. The assessment model proposed by the authors and the data presented in the paper should support similar studies performed elsewhere in the world and help in developing policies for EVs [35].

Further research in this area has been presented in a study concerning a free-floating bike sharing (FFBS) system proposed as a low-emission mode of transport [36]. The authors of the study have pointed out that, even though LCA is a method intended for analysing the environmental impact of FFBS, it rarely takes into account the entire travel chain where intermodal means of transport are used [36]. Therefore, they have proposed an LCA that takes the travel chain into consideration. Their analysis covers the production, use, operation, and end-of-life stages of this solution in Nanjing, China. As part of the study, an online survey was conducted to identify the mechanism of modal change under the impact of FFBS. The results thus obtained show that the majority of respondents have switched to urban cycling from lower-emission transport modes (28.30% from bus, 14.86% from metro, and 33.97% from non-motorised modes) [36]. From such a perspective, cycling in fact increases emissions. What the research further implies based on the current number of FFBS systems and the assumption that each bicycle is used once a day on average is that FFBS should be used for at least 227 days to generate positive environmental benefits [36].

OVfPwSN type cycles are devices which typically feature more electronic systems and accessories than average bicycles, including electric bikes. One should definitely analyse the impact this aspect has on the reduced environmental footprint of bike sharing systems. The authors of another study point out that public bicycles have evolved technologically, moving from smart docking systems to smart bicycle systems, and it is unclear whether the increased use of electronics yields any net environmental benefits [37]. Also noteworthy in this context is the lack of research on this problem using, for example, swarm intelligence algorithms. Therefore, it would be reasonable to assess the impact of incorporating further technologies into bicycles (hardware, software, and means of telecommunication). To this end, the authors have analysed the impacts generated by a smart bicycle using LCA [37], and have highlighted that, during the production phase, smart bicycles generate approx. three times more greenhouse gas emissions compared to docking system smart bicycles as converted into a kilometre travelled over the bicycle lifetime [37]. They have further concluded that, considering individual end point categories of human health, ecosystem, and resources, the environmental impact of smart bicycles is ca. 2.7 times greater [37]. The results imply that, when switching from a smart docking station to a smart bicycle, the number of users must increase by a factor of 1.8 in order to compensate for the augmented environmental impact in terms of greenhouse gas emissions [37]. Further analyses of interest concern the impact of demographic factors on this problem. The authors argue that smart docking stations become preferable at population densities ranging from 1030 to 3100 inhabitants per km^2 (depending on the degree of bicycle-friendliness of given city) [37].

The approach presented in yet another publication on this subject is slightly more general [38]. It compares means of transport in terms of their environmental impact by way of a full-scale life cycle assessment. For that purpose, the EIO-LCA (Economic Input Output Life Cycle Assessment) methodology from Carnegie-Mellon was applied. The study comprised the stages of production, installation, and operation of the relevant infrastructure. All these stages were analysed for the required energy input. What followed was a calculation of the impact of the aforementioned stages on pedestrian, bicycle, and electric traffic for both individual and public transport vehicles. The author's results show that electric bicycles use less than 10% of the energy required to power a saloon car for each mile travelled. Additionally, they emit 90% less pollution per every passenger-mile travelled than a bus (operating off-peak) [38]. With regard to the subject matter of this article, in the context of the OVfPwSN currently being engineered, further intriguing problems may be those related to the tourist traffic on mountain trails whose characteristics make them prone to congestion, and there are such trails in Poland.

Another article provides a discussion on interesting research concerning diverse problems associated with the subject of this paper. The authors point out that, while studying the social problems affecting cyclists is rather common, they believe that studies of the environmental, economic, and vitality related effects of cycling infrastructure have not yet become popular [39]. In their research, the authors compared the expected environmental and economic impacts associated with the construction of a cycle path in Mahasarakham, Thailand. They employed both the LCA and life cycle costing methods to evaluate the cycle path from an environmental and economic perspective [39]. The scope of the study covered processing, acquisition, and transport of raw materials, as well as construction, operation, and disposal. The functional unit envisaged under this study was the use of the cycling traffic lane for at least 1 year. Another aspect examined in the research was the environmental impact of greenhouse gas emissions, otherwise known as *carbon footprint*. According to the authors, had such a cycle path been built, emission of approx. 0.2 million tonnes of carbon dioxide, understood as the carbon footprint equivalent, could have been avoided in 2020 alone. The operation phase is of paramount importance to carbon footprint reduction. Environmental impacts can be diminished by reducing the fuel consumption of cars and motorcycles, where bicycles are used in the infrastructural context of a specific location [39]. Importantly, however, only 26% of road users claim they would be willing to switch to a bicycle if such a dedicated path were built [39].

Figure 1 shows the research problems of relevance to the subject matter at hand, identified on the basis of the foregoing literature review. Even the vehicle structure itself, defined at the design stage, exerts a certain environmental effect. It is the design that essentially determines the vehicle's further environmental impacts. Another factor assumed to condition environmental impact is the need for means of production (land, capital, labour). And it is quite relevant in this respect, since many of the production facilities releasing vehicles of this type are companies built from scratch, meaning that they did not make use of any ready-made (previously installed) production lines.

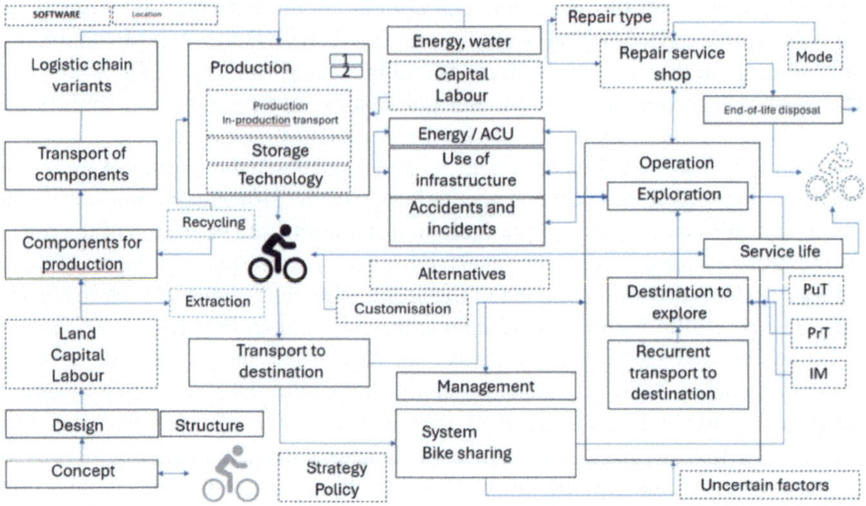

Fig. 1. Identified research problems. Source: Own study based on [6–39].

What matters for the components used in the production of vehicle subassemblies is the share of diverse raw materials, especially those of fossil origin, as well as the share of recycled components. Another environmental impact is related to the transport of components and raw materials for vehicle production. In this regard, there is a variety of logistic chains which have diversified and significant influence on environmental impacts. Some of them are global chains (e.g. those of battery production, lithium, graphite), encumbered by considerable risks (e.g. in cases of armed conflicts, formalisation and implementation of new alternative production technologies). However, the process which can be attributed the greatest environmental impact, as evidenced above with reference to relevant sources, is production. It comprises an abundance of elements that can affect the environment. These include technological processes, assembly, in-production transport, internal transport, storage, packaging, etc., as well as energy and water consumption. What matters at the stages of component sourcing from both remote locations and places in close proximity to the production site (lesser transport-related impact) is the share of recycled components. Every vehicle is parked somewhere at a destination linked with its owner. This means that it requires transport from the production site to that location, which entails a major environmental impact (up to 10%). What may happen next, depending on whether the customer is an individual or a legal entity, is the installation of docking stations for a bike sharing service. An equivalent of this process is the use of a vehicle storage place by an individual (in a garage, under a shelter, etc.). If such a facility is yet to be built, this involves an additional environmental impact. The next phase is the vehicle operation, which ends with disposal, while in between, there are consecutive events that significantly affect the environment. If the vehicle is used outside the place of residence, its use is almost always associated with transport to a destination where a tourist trail or a recreational area will be explored. This form of transport is of a repetitive nature, each time generating a certain carbon footprint, and for its purposes one can move the OVfPwSN using private (PrT) (as long as it provides enough storage space) or public (PuT) means of transport (as long as it offers a suitable connection and provides adequate functionality enabling the vehicle to be carried). It is also possible to reach the tourist trail exploration destination by combining individual and public means of transport, or by adding alternative variants, such as direct use of the OVfPwSN (intermodal transfers: IM=PuT+(PrT)+OVfPwSN). The transport stage is followed by the vehicle operation in mountainous terrain. There are further events associated with the operation that increase the environmental footprint. Firstly, infrastructure is used, which affects the environment (production of infrastructure elements, construction, assembly, installation, replacement, repair, etc.). A mountain trail is not just a natural road. Secondly, vehicle batteries are used, and they also require power supply; it is also frequent that more than one battery is operated throughout a single transfer. Depending on the user's preference, significant wear and tear of successive batteries may be observed. This involves the problem of construction and operation of battery charging infrastructure and battery replacement stations available on mountain and hiking trails. Yet another issue is that of traffic incidents and accidents happening on mountain and hiking trails. They may, for instance, require intervention of Mountain Volunteer Search and Rescue services (Polish GOPR) using a helicopter or other vehicles, thus introducing another considerable environmental burden. And if that entailed long-term rehabilitation, it would mean further environmental impacts that are even difficult to estimate. While being operated on a mountain or hiking trail, vehicle components become worn and damaged, requiring repair or maintenance, which generates more environmental impacts. On top of that, repair and maintenance activities depend on the mode of the vehicle operation. A shared vehicle will, generally, require more action in this regard than a vehicle used individually. And last but not least, the environmental impact of a vehicle is affected by the service life of a specific solution, which depends on the design, components used, mode and frequency of operation, vehicle sharing, etc. In Fig. 1, this is represented by the line connecting the vehicle leaving the manufacturing plant with the end-of-life box. The literature defines the minimum number of operation days required to minimise the environmental impact. Figure 1 contains boxes not connected with the product life line (enveloped in a broken line). One of them is related to the software used for LCA purposes which, as evidenced in the literature, yields divergent ratios (up to 13%). What also matters greatly in terms of environmental impact is the location where the vehicle is used, manufactured, or stored. The same applies to the manufacturer's policies for production, product management, competition, and customer service—both long-term and operational. Policies at large, but also in a more specific dimension, e.g. fiscal policy (vehicles of this type being subsidised) or transport policy (subsidies for the purchase of private means of transport, or for public transport operators), have a significant impact on the environmental effects observed. The quality and method of repair or replacement of parts are also of importance in terms of impact. And finally, there is a number of factors whose function from the LCA point of view is uncertain. These include the method and source of raw material extraction (lithium, titanium, silicon, aluminium, graphite), modal split, weather conditions, etc. Virtually every box shown in Fig. 1 could be expanded or broken down into more detailed components, revealing further factors of relevance to a thorough life cycle analysis. This aspect will be addressed by the authors of this paper.

3 LCA for Off-Road Vehicles for People with Special Needs (OVfPwSN)

The following section discusses specifics of off-road vehicles for people with special needs. A review of selected features of these vehicles shows that they should be treated somewhat differently than conventional bicycles and other single-track vehicles under LCA based analyses. This aspect is particularly valid on account of the differences in the design, transport, use, equipment, and sharing of these vehicles. From the LCA perspective, this translates into an extended scope of analysis.

3.1 Specificity of Off-Road Vehicles for People with Special Needs

In the first place, one must determine how the OVfPwSN type cycles differ from other single-track vehicles, for which the literature provides a relatively extensive LCA related input material [6–39]. For this purpose, references have been made to data from the catalogues of leading companies producing vehicles of this type. However, in their catalogues, the manufacturers of these devices also refer to them as electric bicycles, tricycles, or quadricycles, while they also use the term *vehicle* paired with various adjectives, e.g. *adaptive vehicles*. One should keep in mind that these vehicles are often produced for highly specific applications, while those who sell *regular* bicycles typically describe them using far fewer categories. The OVfPwSN manufacturers, on the other hand, distinguish between a number of relevant categories in product catalogues. One of the terms frequently used in this context is an *adaptive vehicle*. They are somewhat similar to conventional electric bicycles on account of the batteries used. But that is where the similarity ends, since their battery capacities are much larger, allowing for an increased range (>190 km) to be covered with a relatively high weight of the vehicle with the user, reaching up to 250 kg (110 kg is the vehicle alone, such as the Coyote Outrider). Yet the weight of the vehicles currently being manufactured depends on the numerous available variants, starting as low as 10–30 kg in road vehicles [3]. The running range of most vehicles with replaceable batteries exceeds a hundred kilometres. In some applications, sound level is also important, which is why they feature silent electric motors on board. Compared to other off-road vehicles, these are quiet in the immediate vicinity (up to xx meters) [3]. Electric motors are mounted directly on the axle, improving traction, enabling steeper slopes to be negotiated, and providing a higher towing power [3]. Unlike in a conventional cycle, heavy loads are carried directly by the vehicle's structure, instead of the tow bar. The steering system is extremely important in these vehicles, which is why it is often different than those installed in typical single-track vehicles. It is typically customised to the driver's arm and hand grip strength. This allows riders with serious limb injuries to operate the vehicle. Throttle and brake control is done by tilting backwards and forwards. However, some of the most important characteristics of the vehicle are its size and folding capability, enabling it to fit into a typical SUV, and often even into certain hatchback/compact car models [3]. Unlike conventional two-wheelers, these vehicles are capable of carrying a large and heavy cargo directly on the frame structure [3]. They are frequently equipped with fully independent suspension and rear-wheel drive to guarantee traction that is suitable for off-road and especially mountainous conditions [1–4]. They typically feature very wide tyres (in which case the term FAT appears in the vehicle's name). They differ from bicycles in terms of the rider's position, having a different seat which enables hours of exploration of difficult mountain and hiking trails. They are either arm or foot-powered, in various gear ratio combinations, with optional electrically-assisted or fully electric drive [2]. The wheels and full suspension coupling solution makes these vehicles relatively efficient in damping vibrations and compensating uneven terrain. More powerful motors are offered in diverse variants, providing adequate acceleration and increased power [2]. The aforementioned longer running distance can be covered on a single charge or using the battery replacement feature. These vehicles are designed in three- and four-wheeled versions, in diverse configurations (seat, kneeling cycle, rear or front drive axle, different variants of suspension, drivetrain, and steering systems). What makes them stand out against conventional bicycles is the available selection of accessories [1–4]. Such a vehicle can feature special signalling devices, a drone, additional space for the rider's wheelchair (used to navigate through the urban space), an LCD screen, additional batteries, additional lights, sensors not found in conventional single-track vehicles, seat with belt, reverse gear, flag, etc. The above description implies that an OVfPwSN is a completely different means of transport than a bicycle or scooter, even those that come in electric versions. It is by far more technically complex and specialised. It is for this specificity that these vehicles should not be analysed for LCA purposes like ordinary cycles. Their structure is typically made of special alloys, they use different compensation solutions in the steering system, and above all, they feature a seat with a backrest instead of an ordinary bicycle seat, motors that differ in the power range, reaching up to 3 kW/5 kW, and batteries with capacities of up to 1800–6000 Wh. Their typical dimensions are completely different, which strongly conditions their operation; such dimensions may prove somewhat troublesome in urban conditions, while they prove very helpful on difficult trails. And finally, these vehicles feature a wide range of additional equipment, hardly to be found in conventional two-wheelers (the use of a drone on board a OVfPwSN has been described in another article).

Figure 2 highlights selected features and aspects which make OVfPwSN specifically different from conventional single-track vehicles.

3.2 LCA for Off-Road Vehicles for People with Special Needs

The elements relevant to the environmental analysis of single-track vehicles, identified with reference to the above literature review, are as follows: vehicle design, utilisation of ready-to-use production lines, vehicle components, share of fossil raw materials, share of recycled components, transport of components and raw materials, logistic chains, manufacturing process, assembly, in-production transport, storage, packaging, energy and water consumption, parking at the owner's destination, transport from the production facility, installation of docking stations for shared bike systems, storage locations (including depots), operation phase, end-of-life phase, transport to terrain exploration destinations, private transport, public transport, intermodal transport, trail/road infrastructure operation, vehicle batteries, battery power supply, construction/

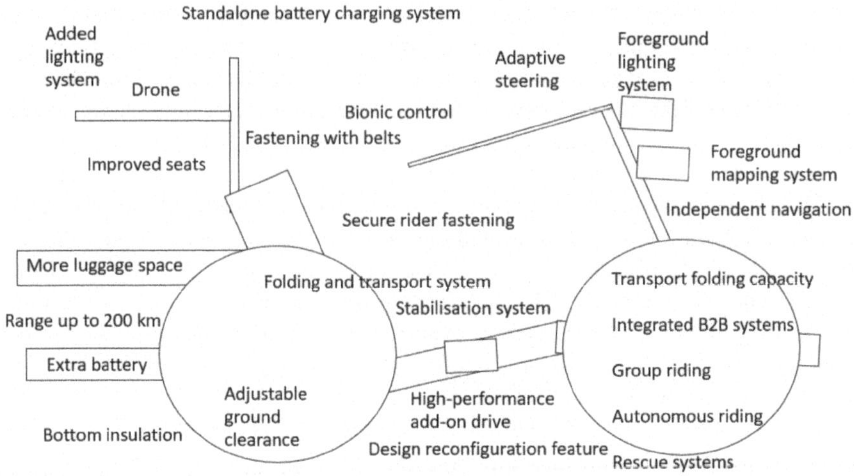

Fig. 2. Specific features of OVfPwSN type vehicles. Source: In-house study.

operation of charging stations, battery replacement, traffic incidents and accidents, wear and tear of components, vehicle sharing, vehicle service life, frequency of use, location of use (local conditions, aggressive environment, temperatures, etc.), transport policy, repair quality and method, raw material extraction method and source, modal split, atmospheric conditions [11–39].

The literature on the subject, covered by the authors' review, refers to some of the foregoing impact factors as quantitative: product transport to the user, production process, operation phase, end-of-life phase, vehicle sharing (US study), vehicle service life (in specific cases), frequency of use, and modal split (US study). Some of the impacts require research extending beyond the life cycle of such a vehicle and performed in a global or local context, and these are: share of fossil raw materials, logistic chains, operation of cycle path infrastructure, transport policy, raw material extraction method and source. Case studies also require meticulous analysis and reference to the local context of the analysis. Therefore, when performing an LCA for OVfPwSN, it may be worthwhile focusing on problems that are relevant but have not been quantified, leaving those of marginal significance aside. Considering the foregoing, an LCA based analysis of OVfPwSN type vehicles should cover such problems as:

- vehicle design, or how to engineer a more environmentally efficient vehicle,
- utilising ready-to-use production lines, or how to manufacture OVfPwSN by making the most of the production capacity at hand (machinery and equipment, entire facilities),
- how to select and design vehicle subassemblies to make the vehicle more efficient in terms of environmental impacts,
- what is the possible and what is the optimal share of recycled components,
- how to organise in-production transport, storage, and packaging, assuming that the production volume of this kind of vehicles is expected to increase dramatically (production line parameters will increase),

- how to organise parking while ensuring minimised environmental impact,
- how and where to install docking stations for sharing systems,
- whether or not, and if so, how to organise OVfPwSN rentals,
- how to manage and plan transport to terrain exploration destinations using various modes of transport: private, public, and intermodal (including traffic using OVfPwSN),
- how to build and operate battery charging stations, especially those located on trails,
- how to minimise the number of incidents and accidents on mountain and hiking trails,
- how does wear and tear of OVfPwSN components proceed,
- how does vehicle sharing affect environmental impact conditions,
- how relevant is the vehicle use location (aggressive environment, temperatures, etc.),
- how does the modal split of the OVfPwSN transport to the destination point affect the place of land exploration, assuming a given modal split on transport service lines?

Each of the aforementioned problem areas represents a subject for a separate project (Fig. 3).

As part of a competition organised by National Centre for Research and Development, entitled *Things are for People*, the Faculty of Transport and Aviation Engineering of the Silesian University of Technology received funding for the following project: *MOUNTAINS WITHOUT BARRIERS—Integrated platform for the planning, organisation and oversight of as well as support for the access to mountain tourism services for people with physical difficulties in functioning, and a special-purpose off-road vehicle*

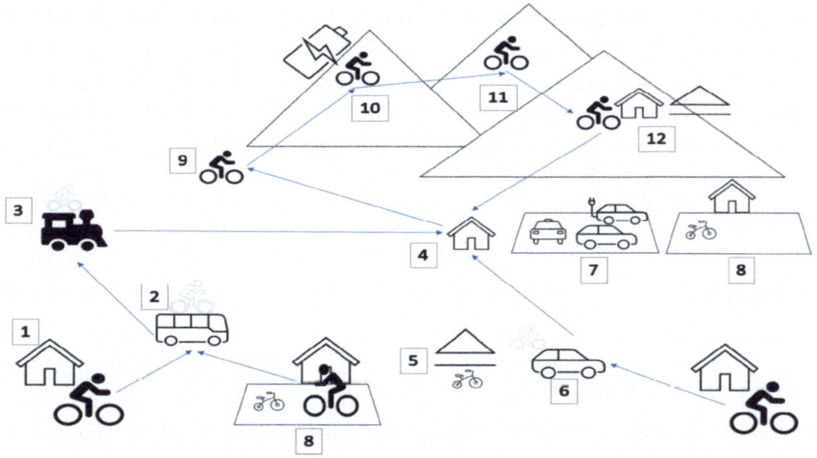

Fig. 3. Schematic diagram of OVfPwSN movement: 1/ home, 2/ means of public transport, 3/ subsequent means of public transport, 4/ exploration start point (bus station, railway station, stop), 5/ OVfPwSN garage or shelter, 6/ individual means of transport, 7/ parking at exploration start point, 8/ OVfPwSN rental shop, 9/ on-trail movement, 10/ on-trail battery charging and replacement station, 11/ on-trail intermediate point, 12/ accommodation base. Source: In-house study.

intended to support the provision of tourism services. In the course of the related studies, specific project premises were defined, the structure of such a vehicle was designed, and a prototype was built. The vehicle engineered at the Department of Transport and Aviation Engineering of the Silesian University of Technology was given the name TRANILOT. The vehicle was built on a four-wheel chassis. Vehicles of a similar design offered by other producers are rather scarce, they are typically intended for highly specific applications, such as hunting or angling, and they are often characterised by a considerable weight. With such parameters, they also offer a significant operating range and admit a high combined driver and cargo weight.

With regard to the foregoing, the project team is preparing to conduct both an LCA and some other research work pertaining to this vehicle after the project in question has been completed. At this stage, the overall body of LCA problems for the said OVfPwSN has been identified. For that reason, the vehicle will be equipped with a system of inertial units and accelerometers for data acquisition, with a view to future LCA analyses. What is also planned under the relevant studies is using other instruments at the research unit's disposal, including 3D scanners, eye tracking instruments, etc. This will comprise an original approach to the LCA research field, previously not encountered in papers on similar subjects, intended specifically for off-road mountain vehicles operated by people with special needs. However, a discussion on these studies exceeds the thematic scope of this article. Further work on the LCA of the vehicle in question will be conducted once the project is completed, financed from the budget of the Faculty of Transport and Aviation Eng. Given the profile of the activities undertaken at the Faculty, the project team is planning to apply for project funding in the following scope:

- Operation of docking stations for bike sharing services in mountainous areas, based on the infrastructure of hiking trails and the local road network,
- Organisation of rental shops and the sharing process for off-road vehicles,
- Adaptation of the collective public transport offering to support the movement of people with special needs who own oversized means of individual transport,
- Design of battery charging and replacement stations based on the infrastructure of tourist trails and the local road network. As extensive as they are, the above topics still do not exhaust the body of problems in question.

4 Conclusion

As the foregoing literature review implies, the overall body of problems of the LCA analysis of both conventional and electrically driven single-track vehicles has been explored relatively thoroughly [11–39]. Based on the review, the main sources of their environmental impact have been identified, including the processes of production, assembly, product packaging, transport of components and raw materials, transport of the finished product to the customer, and vehicle operation. For most of the cases analysed, the production and operation processes have been included among major impacts, each individually exceeding the sum of the environmental impact indices determined for the other identified environmental impact components (e.g. the Amper report).

Several literature sources address specific problems like: bike sharing, identification of problematic subassemblies in the bicycles currently in production, the modal split problem, or the problem of the infrastructure use, etc. Interestingly, the following subassemblies appear to generate the most negative environmental impact (in order of emissions in kg of CO_2 eq.): frame, charger, rims, forks, batteries, crank mechanism.

The authors of this article have specifically highlighted the following problems, often related to the specificity of OVfPwSN:

- vehicle storage, both in the parking area and on the trail, which can increase emissions on account of additional parking infrastructure, parking buildings, lighting infrastructure, sewage infrastructure,
- choice of a passenger vehicle with adequate luggage space, which may increase emissions, since owners of OVfPwSN tend to buy larger (less environmentally friendly) vehicles, while a typical wheelchair for everyday use does not require such storage space, fitting into almost any car,
- multiple transport of the vehicle to the destinations where it is operated/exploration takes place, which usually entails the use of larger individual vehicles or the problem of a choice of means of public transport (while it may even happen that adequate connections are non-existent at all),
- intermodal links: OVfPwSN combined with PrT and PuT means of transport,
- organisation of a network of charging stations and battery replacement stations in mountain areas, partially covering local roads in areas adjacent to these territories,
- organisation of adequate accommodation base for OVfPwSN users and their vehicles,
- security and organisation of mountain and hiking trails in the context of potential traffic incidents and accidents on the trail, although the vehicles in question are more stable in use than conventional bicycles,
- LCA based analysis concerning the conditions in which the vehicle is operated, since even the Polish mountain ranges are rather diversified, enabling traffic of these vehicles, yet under extremely variable weather conditions,
- modal split vis-à-vis the problem of operation of these vehicles in urbanised areas,
- design of the means of public transport vs adaptation to the needs of the transport of OVfPwSN; at the moment, there are virtually no such means; very few transport operators are in disposal of suitable lifts and platforms; fastening elements are missing (as a side note, even users of conventional bicycles often face problems trying to transport their bikes by public means of transport).

The perspective of the problem in question, as presented in this article, as well as its current status and the research completed in this field, all imply that this is in fact a complex issue, comprising numerous secondary problems, extensive enough to justify their independent exploration under separate research projects.

Acknowledgements. The present research has been financed through the National Centre for Research and Development as a part of the competition within the scope of "Things are for people" in project with title: "Integrated platform for planning, organization, supervision and support for the availability of mountain tourism offer for people with difficulties in physical functioning and a specialized off-road vehicle for the implementation of the tourism offer—Mountains without barriers" realized by Silesian University of Technology.

References

1. DEKRES: https://www.dekers.pl/o-nas/. Accessed 29 Jun 2024
2. SPORT-ON: https://sport-on.com/. Accessed 29 Jun 2024
3. OUTRIDER: https://outriderusa.com/. Accessed 29 Jun 2024
4. MOUNTAINTRIKE: https://www.mountaintrike.com/. Accessed 29 Jun 2024
5. GREEN Mountain: https://greenmtnadaptive.org/about/. Accessed 29 Jun 2024
6. Kowalski, Z., Kulczycka, J., Góralczyk, M.: Ekologiczna ocena cyklu życia procesów wytwórczych (LCA). Wydawnictwo Naukowe PWN, Warszawa (2007)
7. Life Cycle Assessment (LCA): A guide to best practice editor(s): Walter Klöpffer, Birgit Grahl. Wiley-VCH Verlag GmbH & Co. KGaA (2014)
8. Venditti, R.A.: Life cycle analysis and sustainability assessment. https://faculty.cnr.ncsu.edu/richardvenditti/wp-content/uploads/sites/24/2018/10/ (2013)
9. Grzesik, K.: Wprowadzenie do oceny cyklu życia (LCA) – nowej techniki w ochronie środowiska. Inżynieria Środowiska. **11**, 101–113 (2006)
10. Weidema, B.P., Cappellaro, F. Carlson, R., Notten, P., Pålsson, A.Ch., Patyk, A., Regalini, E., Sacchetto, F., Scalbi, S.: Procedural guideline for collection, treatment, and quality documentation of LCA data. https://lca-net.com/files/V2004_ProceduralLCA.pdf (2004)
11. Coelho, M., Diogo, A.: Cycling mobility – a life cycle assessment based approach. Transp. Res. Procedia. **10**, 443–451 (2015). https://doi.org/10.1016/j.trpro.2015.09.094
12. Environmental management — Life cycle assessment — Principles and framework, ISO 14040:2006. https://www.iso.org/standard/37456.html/. Accessed 29 Jun 2024
13. Environmental management — Life cycle assessment — Requirements and guidelines, ISO 14044:2006. https://www.iso.org/standard/38498.html. Accessed 29 Jun 2024
14. ISO 14000 family Environmental management. https://www.iso.org/standards/popular/iso-14000-family/. Accessed 29 Jun 2024
15. Stilo, L., Segura, V.D., Lugo, H., Conway, P., West, A.: Electric bicycles, next generation low carbon transport systems: a survey. Transp. Res. Interdiscipl. Perspect. **10**, 100347 (2012). https://doi.org/10.1016/j.trip.2021.100347
16. Cherry, C., Weinert, J., Xinmiao, Y.: Comparative environmental impacts of electric bikes in China. Transp. Res. Part D: Transp. Environ. **14**, 281–290 (2009). https://doi.org/10.1016/j.trd.2008.11.003
17. Calão, J., Marques, D., Completo, A., Coelho, M.: Life cycle thinking approach applied to a novel micromobility vehicle. Transp. Res. Rec.: J. Transp. Res. Board. **2676**, 036119812210846 (2022). https://doi.org/10.1177/03611981221084692
18. Kazmaier, M., Taefi, T., Hettesheimer, T.: Techno-economical and ecological potential of electric scooters: a life cycle analysis. Eur. J. Transp. Infrastruct. Res. **20**, 233–251 (2020). https://doi.org/10.18757/ejtir.2020.20.4.4912

19. Chen, Y., Zeng, D., Deveci, M., Coffman, D.: Life cycle analysis of bike sharing systems: a case study of Washington D.C. Environ. Impact Assess. Rev. **106** (2024). https://doi.org/10.1016/j.eiar.2024.107455
20. Life cycle analysis Tanja Tanskanen October 2022. https://www.nimms-rad.de/news/wp-content/uploads/2015/06/Ampler-LCA-Report-EN.pdf
21. GREENDELTA: https://www.greendelta.com/software/. Accessed 29 Jun 2024
22. Wang, S., Wang, H., Xie, P., Chen, X.: Life-cycle assessment of carbon footprint of bike-share and bus systems in campus transit. Sustainability. **13**(1), 158 (2021). https://doi.org/10.3390/su13010158. Accessed 29 Jun 2024
23. Luna, J.: Bicycle life cycle. https://www.designlife-cycle.com/bicycle (2016). Accessed 29 Jun 2024
24. Breene, S.: Beginner's guide to the most common bike repairs. Greatist, 9 Oct 2012. Web. 14 Mar 2016. http://greatist.com/fitness/bike-repair-maintenance-guide (2012)
25. Johnson, R., Kodama, A., Willensky, R.: The complete impact of bicycle use: analyzing the environmental impact and initiative of the bicycle industry. Master's project, Duke University. https://hdl.handle.net/10161/8483 (2014)
26. Roy, P.: Environmental Life cycle assessment of family/cargo e-bike: a cleaner solution for urban mobility. Master's thesis, University of Graz. https://unipub.uni-graz.at/obvugrhs/content/titleinfo/8285669/full.pdf (2022)
27. SIMAPRO: https://simapro.com/. Accessed 29 Jun 2024
28. Tayyab, M., Singari, R.M., Sathik, P.M.: Evaluating the environmental impact of a bicycle: a life cycle assessment with a new green product design framework educational administration: Theory Pract. 2024, 30(5), 3823-3831 (2024)
29. Roy, P., Miah, M.D., Zafar, M.T.: Environmental impacts of bicycle production in Bangladesh: a cradle-to-grave life cycle assessment approach. SN Appl. Sci. **1**(7), 700 (2019). https://doi.org/10.1007/s42452-019-0721-z
30. Montoya-Torres, J., Akizu-Gardoki, O., Iturrondobeitia, M.: Life cycle assessment via software tools: result analysis in a bicycle study. In: Manchado del Val, C., Suffo Pino, M., Miralbes Buil, R., Moreno Sánchez, D., Moreno Nieto, D. (eds.) Advances in Design Engineering IV. INGEGRAF 2023 Lecture Notes in Mechanical Engineering. Springer, Cham (2024). https://doi.org/10.1007/978-3-031-51623-8_75
31. OPENLCA: https://www.openlca.org/openlca/. Accessed 29 Jun 2024
32. SOLIDWORKS: Accessed 29 Jun 2024
33. Kasmer, M.Y.I., Fahrul, H.: Life Cycle assessment of a bicycle using solidworks sustainability. http://publisher.uthm.edu.my/periodicals/index.php/rpmme
34. Guangnian, X., Qiongwen, L., Anning, N., Zhang, C.: Research on carbon emissions of public bikes based on the life cycle theory. Transp. Lett. **15**(4), 278–295 (2022). https://doi.org/10.1080/19427867.2022.2123142
35. Huang, Y., Jiang, L., Chen, H., et al.: Comparative life cycle assessment of electric bikes for commuting in the UK. Transp. Res. Part D: Transp. Environ. **105**, 103213. ISSN 1361-9209 (2022)
36. Lai, R., Ma, X., Zhang, F., Ji, Y.: Life cycle assessment of free-floating bike sharing on greenhouse gas emissions: a case study in Nanjing, China. Appl. Sci. **11**(23), 11307 (2021). https://doi.org/10.3390/app112311307
37. Bonilla-Alicea, R.J., Watson, B.C., Shen, Z., Tamayo, L., Telenko, C.: Life cycle assessment to quantify the impact of technology improvements in bike-sharing systems. J. Ind. Ecol. **24**, 138–148 (2020). https://doi.org/10.1111/jiec.12860

38. Shreya, D.: Life Cycle Assessment of Transportation Options for Commuters. Massachusetts Institute of Technology (MIT). https://teslaclubsweden.se/wp-content/uploads/2015/06/Pietzo_LCAwhitepaper.pdf (2010)
39. Prasara, J., Bridhikitti, A.: Carbon footprint and cost analysis of a bicycle lane in a municipality. Glob. J. Environ. Sci. Manage. **8**(2), 197–208 (2022)

Request Tram Stops—Analysis of Implementation Possibilities in the Selected Tram System

Adam Budziński[1], Mateusz Rydlewski[1](✉), and Agnieszka Tubis[1]

[1] Wroclaw University of Science and Technology, Wroclaw, Poland
255388@student.pwr.edu.pl; mateusz.rydlewski@pwr.edu.pl; agnieszka.tubis@pwr.edu.pl

Abstract. Request stop systems are a popular solution in the context of public bus transport. In the case of tram systems, this solution is definitely less popular. The article presents the advantages of this type of solutions and describes the potential threats resulting from the operation of request tram stops. Functioning solutions for request tram stops in Poland were presented. The conducted research included measuring the number of vehicle stops at stops in one of the existing tram transport networks. This study was repeated on selected tram routes in a city where this type of stops do not exist—in Wrocław. The obtained results indicate that it is reasonable to consider the possibility of implementing this solution.

Keywords: Tram system · Tram stops · Request stops

1 Introduction

Public transport systems in Poland are most often based on bus lines, which are a common solution operating in every city. There are tram systems in 15 urban centers in Poland, and the metro is only available in the country's capital. An integral part of the functioning of public transport systems are stops where passenger exchange takes place. Research [1] indicates that in Polish conditions, for trams the time spent on stops makes up the remarkable amount of 30% of the total time of tram line operation. Such a high share of waiting time at stops in relation to the entire transport process creates the potential to analyze tram stops in the context of the functioning of this transport system.

Regardless of technical and infrastructural aspects, stops can be divided into two types: permanent and request. The difference between them is that a vehicle serving a permanent stop will always stop there, and in the case of a request stop, only if the passenger who is in the vehicle or waiting at the stop so wishes. The use of request stops is common in bus lines and much less common in tram lines. In selected systems, request stops are also used in rail transport [2]. The less popular use of request stops in tram systems is probably related to the characteristics of tram traffic, which is pointed out by

G. Sierpiński et al. (Eds.): TSTP 2024, LNNS 1257, pp. 129–144, 2024.
https://doi.org/10.1007/978-3-031-81777-9_9

the author [3]. The following aspects are noted as problems with implementing this type of stop for the tram network [4, 5]:

- impossibility of two trams passing,
- introducing disruptions to the timetable,
- using the alarm system to inform the tram driver about his intention to leave the tram,
- problems of blind people and people who do not know the area in which they are located,
- problems resulting from insufficient visibility,
- specifics of the tram braking process.

The advantages of request stops include [4, 5]:

- lower fuel consumption (or electricity in the case of trams),
- lower use of rolling stock (globally),
- shortening travel times between subsequent stops,
- increased driving comfort due to reduced number of accelerations and decelerations (braking),
- increasing the operating speed,
- reduction of energy losses—which is both an environmental and economic benefit due to the reduction of carrier costs.

The braking distance of a tram is much longer than that of buses. These differences are presented in Table 1. The terms used there should be understood as follows [3]: stopping distance (SD)—the distance expressed in meters determining the distance covered from the moment of braking, and breaking distance (BD)—the distance expressed in meters, which is the sum of the stopping distance and the distance traveled during the driver's reaction time to a situation requiring braking (so-called braking delay).

The aspect related to stopping and braking trams is an area of research that concerns optimization of the operation of traffic lights. Research results [6, 7] indicate the benefits of reducing the number of tram stops at intersections with traffic lights. This also has a direct impact on request stops, where stopping takes place only when there is an actual need on the part of the passenger.

The implementation of request stops may have a significant impact on aspects related to the design of timetables, as pointed out by the authors of the study [8]. Lack of stops at requested stops may result in overspeeding of vehicles at subsequent stops or delays in the event of stops at the request of passengers. Research in the Spanish city of Santander [9] relating to the optimization of the distance between stops indicates the importance of ensuring accessibility to stops, which, along with decreasing demand, should determine longer distances between stops. Increasing the distances between

Table 1. Stopping and breaking distance for buses and trams, source: own work based on [3].

	BUS		TRAM	
SPEED	BD	SD	BD	SD
20 km/h	3.6	9.5	7.7	13.2
50 km/h	18.9	33.8	49	60

stops increases the commercial speed and, combined with the offer of request stops, does not reduce access to public transport services. In [10] the aspect of request stops in the context of water transport was discussed. The results of the conducted analyzes indicate the benefits of implementing an request stop solution, which is located on the route between the initial and final point of the transport service. In addition to operational cost savings, increased passenger satisfaction has also been demonstrated as travel times on the main section have been reduced. Reducing the costs of providing services is an important aspect that is particularly important for public transport operators. As indicated by [11], cost reduction should be achieved while maintaining the offer and commuting options, especially in areas with lower traffic potential.

The author of research on request stops [4] proposed a method for determining the status of stops in urban public transport. It is based on two factors:

- stops factor KZ—which is the quotient of the number of stops and the total sample (number of stops),
- service factor KO—which is the quotient of the number of stops with at least one door opened and the total sample (number of trips).

In a computational example presenting the values of individual factors for the studied area, the author [4] proposed the classification of stops depending on the value of the service factor. For values above 80%, it is recommended to operate the stop as a permanent stop. Stops with a ratio below 50% should be request stops. In turn, values between 50% and 80% constitute the basis for observations, but in the author's opinion they can be used both as constants and on demand, depending on additional circumstances and analyses. Taking into account the similarity of the described method with the subject of the research, the factors proposed and described above were used in the further part of this chapter, describing our own research.

2 Tram Systems with Request Stops in Poland

Request stops in Poland are primarily used in bus networks, which are typically found in less urbanized zones. The ease of implementing the request stop system for buses is due to the greater mobility of these vehicles in terms of acceleration and braking, compared to other rail-based modes of transport.

In Poland, there are three tram networks where trams stop at request stops upon a passenger's request. These tram networks are located in:

- Metropolis GZM,
- Kraków,
- Szczecin.

In recent years, there existed a fourth tram network with a request stop in Poland. These stop, the only one in the network, was located at the Metro Młociny stop cluster for tram line number 2. However, the status of this stop was changed to a regular stop in 2023 [12].

The remaining networks are characterized by having only regular stops in their systems, meaning that vehicles are required to stop at these locations.

2.1 Metropolis GZM

The tram network located in southern Poland in the Silesian Voivodeship is one of two that serves more than one town. These towns are: Bytom, Będzin, Chorzów, Czeladź, Dąbrowa Górnicza, Gliwice, Katowice, Mysłowice, Siemianowice Śląskie, Sosnowiec, Świętochłowice, Ruda Śląska and Zabrze.

Due to its intercity nature, a characteristic feature of such a network can be its coverage of areas of undeveloped land between cities. An example is the tram line between Katowice and Mysłowice, or the line between Bytom and Zabrze. Many request stops are located near former industrial centers (e.g., Szombierki Młyn, Biskupice Ziemka) or sparsely populated areas (e.g., Zabrze Hagera) [13].

The town where request stops have been implemented in tram transportation are Bytom, Ruda Śląska and Zabrze.

The first information available on the Internet about the operation of request stops dates back to a diagram from September 1, 2000. In [14] archival Internet timetables available for a similar period (2000–2001) did not include any information indicating a different status for stops other than regular stops, despite appropriate information on the diagram. The first information in [14, 15] Internet timetables about the status of a stop appeared in the second half of 2002. The number of stops that had the status of a request stop was fewer than today. Over the next 8 years, significant changes in the status of stops occurred, as evident in Table 2. This was influenced, among other factors, by economic changes (such as mine closures) and changes in the tram network (such as temporary route closures or eliminations). Since 2010, two additional stops have been added to the list of request stops: [15] Dąbrowa Miejska Węglowa and [17] Biskupice Schronisko. The list of request stops over the years has been documented in Table 2, utilizing [14, 15] archival sources to analyze their status from 2000 to the present day.

Based on the analysis of data presented in Table 2, it can be inferred that the variability in the status of request stops in metropolitan tram communication has been minimal since 2010. The recent changes from regular stops to request stops affected the stops at (Bytom) Łagiewniki Szyby Rycerskie, which was reinstated after several years of closure, and Dąbrowa Miejska Węglowa, which [16] initially opened on October 30, 2009, as a regular stop but had its status changed to a request stop 6 months later [14, 15]. The most recent modification to the status of an existing stop pertained to Czeladź Strefa Ekonomiczna in 2020. In 2024, a new stop, Biskupice Schronisko, [17] was inaugurated, marking it as the first new stop with request status since 2010.

For this tram network, it is characteristic to lack a unified system for signaling the intention to leaving at request stops. In most vehicles, there is an information advising passengers to press the alarm signal to notify the tram driver of the need of stop at a request stop—examples are illustrated in Fig. 1, depicting this procedure for two different tram types. Additionally, there are trams where no indication prompts the use of an alarm signal. In such cases, a stopping request system similar to those used on buses is enforced within these vehicles.

Table 2. Table of stops that have had request stop status since 2000, source: own work based on [13–15].

(1)	Name of the stop	2000	2002	2006	2008	From 2010
B	Piłkarska[a]	OD				
B	Odrzańska	OD				
BB	Szkoła	DNE	OD			
BDM	Węglowa	DNE				OD[b]
BDM	*Dąbrowa Miejska*	OD				DNE[c]
BŁ	Gojny	RS	OD	*EX*		OD
BŁ	Szyby Rycerskie	OD		*EX*		OD
BR	*Droga do Szybu Lompy*	OD		EX		
BS	Młyn[d]	RS	OD			
Cz	Strefa Ekonomiczna[e]	OD				RS[e]
PBŚ	*Mijanka*	RS	OD	EX		
PDW	*Dom Kultury*[f]	OD		EX		
RŚZ	Ceramiczna[g]	RS	OD			
SK	Ośrodek Opiekuńczy	OD	RS			
W	*Proletariatu*	OD	RS	EX		
Z	Hagera	RS	OD			
ZB	Ziemska[h]	RS		OD		
ZB	Schronisko	DNE				OD

Meaning of abbreviations: OD—request stop (on-demand), RS—regular stop, DNE—stop doesn't exist, EX—route closed or decommissione.

(1) Town and district: B—Bytom (districts; BB—Bobrek, BDM—Dąbrowa Miejska, BŁ—Łagiewniki, BR—Rozbark, BS—Szombierki), Cz—Czeladź, P—Piekary Śląskie (BŚ—Brzezinka Średnia, DW—Dąbrówka Wielka), RŚ—Ruda Śląska (Z—Zgoda), S—Sosnowiec (K—Klimontów), W—Wojkowice, Z—Zabrze (B—Biskupice).

[a]In the year 2000, it was named "Ogródki Działkowe".

[b]The stop was opened in October 2009, initially as a regular stop. The change of status to a request stop occurred in early April 2010.

[c]The stop was decommissioned in 2023.

[d]Until 2005 as Kolonia Zgorzelec.

[e]Until 2020 as Wodociągi.

[f]Until 2005 as Zagrodnicza.

[g]Until 2005 as Bykowina Porcelanka.

[h]Until 2006 as Kopalnia Pstrowski.

2.2 Kraków

Another tram network where request stops are in operation is located in Kraków. A source [18] suggests that the system of stops where vehicles stop upon passenger request may have been in place as early as 1992. In the case of the online schedule [15], information regarding the stop status appeared around mid-2006. This applies to both tram and bus routes. No information has been found to ascertain whether this constituted a reintroduction of the system or merely an expansion of information for passengers seeking details online.

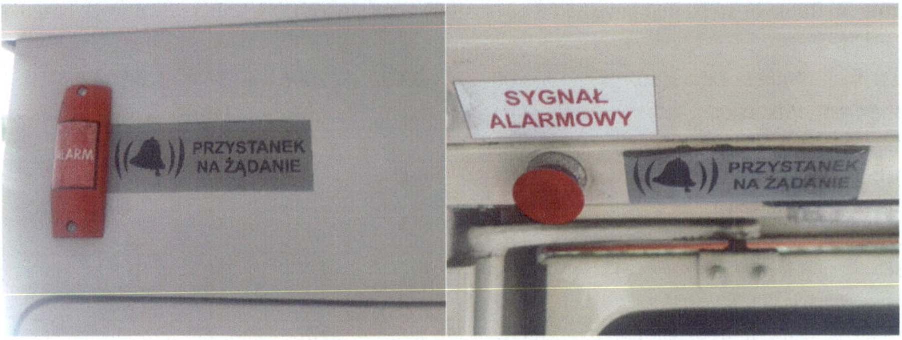

Fig. 1. Button for the request stop signal. On the left, in the Moderus Beta vehicle, and on the right, in the Konstal 105Na vehicle, source: own work.

Fig. 2. A tram stop with different statuses for different lines, source: [20].

According to online sources [20], a photograph suggests that at a certain period, tram stops may have had different statuses for each tram line, as illustrated in Fig. 2. During the peak period [15], the number of stops with optional stopping at the request of a passenger was 15, spanning from 2015 to 2020. Currently, there are eight such stops in Kraków, with the majority located along the route to Pleszów starting from the "Kopiec Wandy" stop [19].

Unlike in GZM, in Kraków, there are regular changes to the status of tram stops. One example is the modification made in December 2023, where the "Muzeum Fotografii" stop was changed to a regular stop, and also the change of status for the "Łagiewniki ZUS" stop from regular to request stop in 2022 [21].

2.3 Szczecin

The third tram network in Poland where request stops are implemented is in Szczecin, where there are 13 stops with such status [22]. Compared to other networks, request stops in Szczecin are a relatively new solution, as the first such stop appeared in the

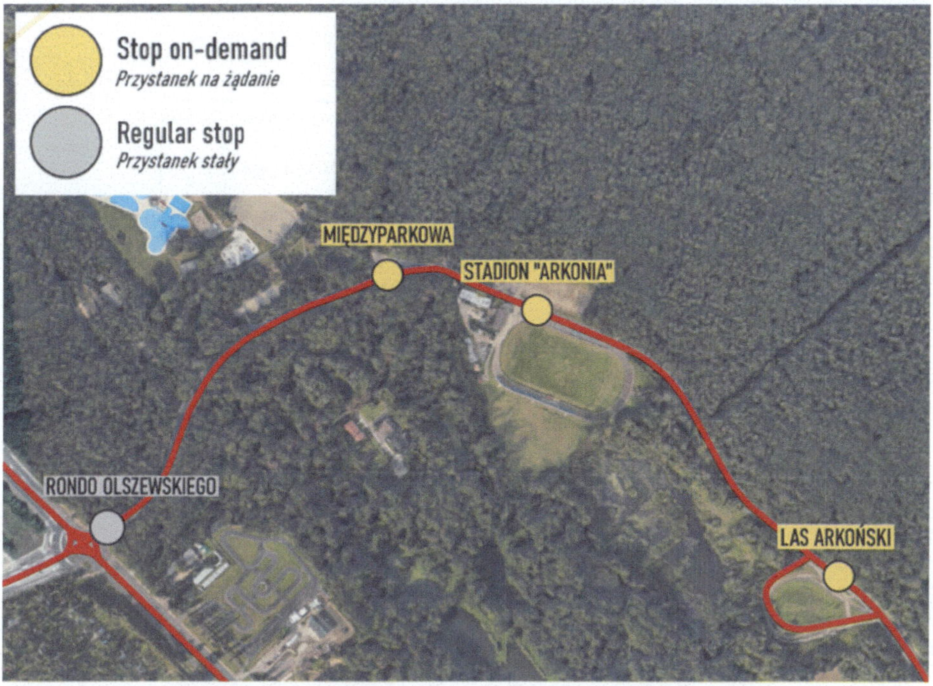

Fig. 3. A map of the tram route from Las Akroński to Rondo Olszewskiego with highlighted stops according to their categories, source: own work based on Google Maps.

schedule in 2015. These stops primarily operate along the route of the Szczeciński Szybki Tramwaj (SST, in English Szczecin Fast Tram), as well as on tram lines to Głębokie and along the Las Arkoński to Rondo Olszewskiego route, including the Bulwar Piastowski stop. Both the SST and the Las Arkoński to Rondo Olszewskiego route are recent constructions, suggesting potential for request stops on new routes. This indicates that stop spacing need not be extensive, even if demand for stops in specific locations is lower. Figure 3 shows an example of the new route, highlighting its path and stops. Attractions along this route likely include recreational facilities such as a stadium, swimming pool, and forested areas.

3 Research Methodology

Achieving the goal of assessing the feasibility of implementing a system of request tram stops in Wrocław included six consecutive steps, which are presented in Fig. 4. Step 1—identification of tram system with request stops was described in chapter 'Choosing of Service Strategy for Cargo Owners in Regional Supply Chains'. Step 2—identification a leading request stop system, indicates that the leader of the three systems presented is the one located in the Metropolis GZM. The rationale for choosing this system has been described above. The next steps refer to the practical parts in

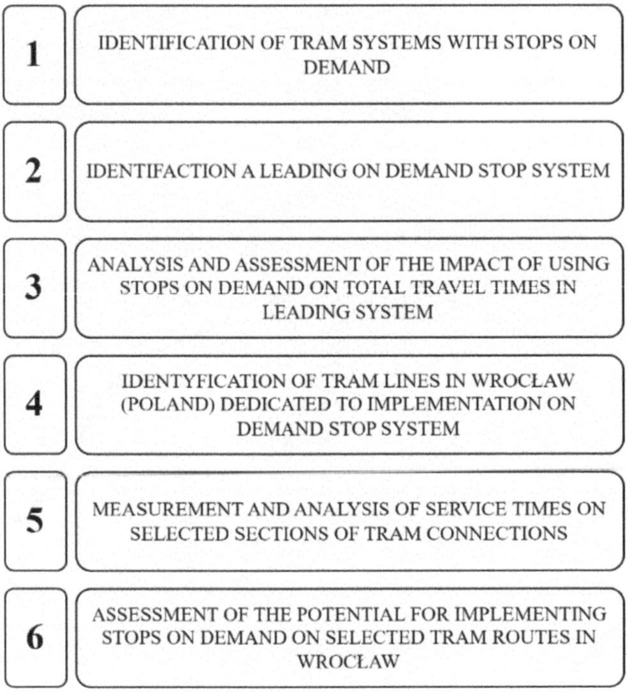

Fig. 4. List of research steps, source: own work.

which field research is carried out, the method and scope of which is described later in this chapter. The scope of the research also includes the tram system in Wrocław in order to verify the feasibility of implementing this type of solution.

The research was conducted between April 25th and June 1st, 2024, between the hours of 10 AM and 8 PM. A total of 64 segment measurements were taken, encompassing 365 measurements related to tram stops. The study was carried out for tram network featuring request stops and for network where all stops have a regular status. Specifically, these are the networks in the Metropolis GZM and in Wrocław. The choice of the Metropolis GZM was driven by the fact that the tram network in this area has a long tradition of tram stops with request status. Their long-standing presence in the studied area allows us to conclude with a high degree of certainty that this solution is well-known among local users. Consequently, it is highly probable that the absence of a tram stop was not due to a passenger error in failing to signal the need for the tram to stop at the stop.

The research was conducted on the following routes:

1. Metropolis GZM
    (a) Zabrze Prof. Religi—Biskupice Pętla—Bobrek Ratusz
    (b) Bytom Olimpijska—Dąbrowa Miejska Centrum Handlowe

2. Wrocław
    (a) Kochanowskiego—Sępolno
    (b) Pilczyce—Leśnica

The selection of routes in Wrocław, where there is no system of request tram stops, resulted from knowledge of the layout and operation of the tram system in Wrocław. The selected sections constitute the outer parts of the tram network and are characterized by reduced passenger traffic, often only at peak times. Additionally, in the case of section a (Kochanowskiego—Sępolno), the argument for selecting this route is also the aspect related to the lack of attractors and traffic generators at selected tram stops. In the case of this route, some stops are located next to a park and in scattered buildings with a predominant character of single-family buildings.

The research conducted in this study primarily aims to assess the temporal and passenger potential of request stops. The parameters defining a request stop are determined by:

- Proportion of stops with zero passenger exchange among all stops,
- Proportion of stops with passenger exchange of up to five boarding and five alighting passengers,
- On the measured routes, the following measurements were conducted:
- Numbers of passengers boarding and alighting at the stop,
- Door opening time,
- Door closing time,
- Departure time from the stop or passage time in case no stop occurred.

3.1 Method of Measuring the Number of Passengers

The measurements were carried out in a traditional way and consisted in determining the number of passengers who took part in the passenger exchange. The selection of the data collection method corresponded to the characteristics of the objects on which the research was conducted. Due to the specificity of request stops and, consequently, the expected small volumes of passenger exchange, the selected method is sufficient. In cases where the number of people getting on or off exceeded 10 people, the exact number of people was not recorded. This means that for measurements in which there was a large passenger exchange, the result was considered to be out of range. If the measurements concerned a route on which one of the stops was the starting or ending stop, the filling level of the vehicles was also recorded.

3.2 Method of Time Measuring

The time measurement and recording of events aimed to determine the time differences between non-stop passages and stops. The timing started at the first stop of the route under investigation, followed by recording times corresponding to subsequent stages of the journeys.

4 Study of Travel Time and Passenger Exchange

In the subsequent part of this chapter, the results of the time and passenger exchange measurements in tram systems in Metropolis GZM (Subsect. 4.1) and in Wrocław (Subsect. 4.2). In the first case, the research was conducted at selected tram stops with request status. In Wrocław, the research covered sections of the tram network where potential passenger exchange is low and request stops may be justified for use and application.

4.1 Study of Travel Time and Passenger Exchange for a Tram Network with Request Stops

The measurements were conducted on two tram routes that have request stops. The aim of this study is to demonstrate the differences in travel time between stopping and non-stopping at a tram stop, and to calculate the stopping ratio within the stop set. The results of the measurements for the request stops in the Metropolis GZM are presented in Table 3. The stop factor *KZ* was calculated according to the method presented in the description of the method for determining the status of stops in the public transport network [4].

The measured stops, visible in Table 3, were request stops at the time of measurement. The indicators were shown, indicating the number of trips without passenger exchange, the percentage of trips through the stop with minimal passengers, and the overall stopping coefficient. Stops for which the sum of the stop factor KZ and the indicator of no passenger exchange exceeded 100% had at least one extra stop at a request status stop.

Based on the justification of maintaining or changing the stop status using the stopping coefficient method, which identified three groups of analyzed stops, the suggested division shown in Table 4 allocates stops to specific groups [4].

Based on the route from Bytom Olimpijska to Dąbrowa Miejska Centrum Handlowe, auxiliary calculations were conducted to determine the impact of non-stop stops on travel time. Out of 48 passages through stops, 36 passages were for request stops, among which 18 were non-stop.

Table 3. Request stops on the surveyed routes along with the obtained indicators, source: own work.

Tram stop		No passenger interchange	Exchange of up to (+ or −) 5 passengers	Stop factor *KZ*
Z	Hagera	11.11%	88.89%	88.89%
ZB	Ziemska	44.44%	55.55%	55.56%
ZB	Schronisko	100.00%	0.00%	0.00%
BB	Szkoła	66.67%	33.33%	33.33%
B	Piłkarska	41.67%	58.33%	66.67%
B	Odrzańska	66.67%	66.67%	33.33%
BDM	Węglowa	58.33%	50.00%	50.00%

Table notation: + indicates boarding passengers, − indicates alighting passengers.

Table 4. Classification of examined stops in Metropolis GZM according to categories, source: own work based on [4].

Proposed stop status	Classification criteria (service factor KO)	Stops classified into this group
Regular	Above 80%	Zabrze Hagera
Regular or request	Between 50% and 80%	Biskupice Ziemska, Bytom Piłkarska, Dąbrowa Miejska Węglowa
Request	Below 50%	Biskupice Schronisko, Bobrek Szkoła, Bytom Odrzańska

Table 5. Travel times between stops depending on the number of skipped stops along the route—towards Bytom Olimpijska, source: own work.

End stop of the surveyed route		*i*	Average travel time for cases		
			No omissions	One omission	Two omissions
Starting stop: BDM Leśna					
BDM	Węglowa	1	**01:05**	00:50	X
B	Odrzańska	2	**02:23**	01:51	X
B	Piłkarska	3	**03:24**	02:52	02:39
B	Olimpijska	4	**04:24**	03:56	03:24
Starting stop: BDM Węglowa					
B	Odrzańska	1	**00:49**	00:47	X
B	Piłkarska	2	**01:50**	**01:50**	01:26
B	Olimpijska	3	02:50	**02:54**	02:11

Calculations of travel time were conducted between the departure time from the i-th preceding stop (the number of stops required to travel from the initial stop to the current one) and the stop being measured. The research results for the direction towards Bytom Olimpijska are presented in Table 5, and the results for the route towards Dąbrowa Miejska are presented in Table 6.

Based on the above tables, it can be observed that in most cases, the longest travel time occurred when no stops were skipped along the route. The longest journeys are highlighted in bold. An exception is noted for the travel time between Dąbrowa Miejska Węglowa and Bytom Olimpijska, where the longest time was measured with one stop skipped. However, this is due to one significantly outlier result being recorded.

The obtained results allow for estimating average values for the time savings that can be achieved in situations where the tram does not stop at a station. Shortening travel times can in turn impact the management of the tram network, including fleet management and capacity at tram loops. In some cases, saving a few minutes on the journey of a particular line can reduce the need for rolling stock, directly affecting service costs without compromising passenger service quality. However, each of these solutions should be considered individually and based on the local conditions of tram system operation.

Table 6. Travel times between stops depending on the number of skipped stops along the route—towards Dąbrowa Miejska, source: own work.

End stop of the surveyed route		*i*	Average travel time for cases			
			No omissions	One omission	Two omissions	Three omissions
Starting stop: B Olimpijska						
B	Piłkarska	1	**00:47**	00:41	X	X
B	Odrzańska	2	**01:50**	01:47	01:29	X
BDM	Węglowa	3	X	**02:43**	X	02:16
BDM	Leśna	4	X	**03:45**	X	03:25
Starting stop: B Piłkarska						
B	Odrzańska	1	**00:52**	00:50	00:48	X
BDM	Węglowa	2	X	**01:46**	01:35	X
BDM	Leśna	3	X	**02:47**	X	02:44

4.2 Study of Passenger Interchange at Tram Stops in a Network Without Request Stops

The study was conducted on two tram routes in Wrocław, covering the following routes:

1. Kochanowskiego—Sępolno, which includes five intermediate stops and terminates at Sępolno terminus.
2. Pilczyce—Leśnica, which includes nine intermediate stops and terminates at Leśnica terminus.

The aim of the study is to assess the feasibility of introducing an request stop system based on the number of stops that could potentially adopt this system. The list of stop locations on the above routes is provided in Table 7, distinguishing stops by the proportion of stops without passenger exchange, with passenger exchange involving up to five passengers, and exchanges involving more than five boarding or alighting passengers. In contrast to the study in Subsect. 4.1, the last parameter (i.e., exchange involving more than five boarding or alighting passengers) is included. This is due to observed passenger exchanges at stops exceeding the assumed values. The stop factor *KZ* was not considered, as all stops in the Wrocław tram network require mandatory stopping.

Similarly to the previous study, based on this, a decision was made to apply a similar classification of stops according to potential status. The classification and assignment of stops to groups have been highlighted in Table 8.

Based on this, it can be inferred that out of the 14 stops surveyed, only 1 clearly indicates the justification for optional stopping. This stop is "Karłowicza". However, considering stops with a questionable status, their number increases to six. When divided by route, we can observe that both routes have a total of three stops each that could be considered as request stops. This division is precisely shown in Table 9, which presents the breakdown of stops by route.

In light of the above results, it can be concluded that there is potential in Wrocław to implement an request stop system in the tram network. While only one stop ("Karłowicza") clearly indicates this, there are five additional stops ("Tarczyński Arena

Table 7. Stops measured on the surveyed routes in Wrocław, source: own work.

Tram stop	No passenger interchange	Exchange of up to (+ or −) 5 passengers	Exchange of more than (+ or −) 5 passengers
Tarczyński Arena (Lotnicza)	28.57%	64.29%	7.14%
Glinianki	35.71%	64.29%	0.00%
Al. Architektów	0.00%	92.86%	7.14%
Grabowa	14.29%	78.57%	7.14%
Kosmonautów Szpital	0.00%	50.00%	50.00%
Kamiennogórska	14.29%	71.43%	14.29%
Złotnicka	0.00%	35.71%	64.29%
Wschowska	7.14%	85.71%	7.14%
Jeleniogórska	42.86%	57.14%	0.00%
Chopina	15.38%	84.62%	0.00%
Karłowicza	61.54%	38.46%	0.00%
Stadion Olimpijski	30.77%	61.54%	7.69%
8 Maja	26.92%	69.23%	3.85%
Godebskiego (AWF)	15.38%	61.54%	23.08%

Table notation: + indicates boarding passengers, − indicates alighting passengers.

Table 8. Classification of examined stops in Wrocław according to categories, source: own work based on [4].

Proposed stop status	Classification criteria (service factor KO)	Stops classified into this group
Regular	Above 80%	Al. Architektów, Grabowa, Kosmonautów Szpital, Kamiennogórska, Złotnicka, Wschowska, Chopina, Godebskiego (AWF)
Regular or request	Between 50% and 80%	Tarczyński Arena (Lotnicza), Glinianki, Jeleniogórska, Stadion Olimpijski, 8 Maja
Request	Below 50%	Karłowicza

Table 9. Division of stops into categories described above by route, source: own work.

Route	Number of regular stop	Number of stops with questionable status	Number of potential request stops
Kochanowskiego—Sępolno	2	2	1
Pilczyce—Leśnica	6	3	0

(Lotnicza)", "Glinianki", "Jeleniogórska", "Stadion Olimpijski", "8 Maja") that could be considered. It is noteworthy that at least two of these stops are located near venues hosting large events ("Stadion Olimpijski" and "Tarczyński Arena (Lotnicza)"). This means that during such events, these stops are very likely to be used by passengers.

Table 10. Average stop durations at surveyed tram stops in Wrocław, source: own work.

Category of stop	Average stop duration [s]
No passenger interchange	7.25
Exchange of up to (+ or −) 5 passengers	12.94
All stops	12.17

Based on the measurement of tram stop durations (Table 10), it was found that the average duration of tram stop without passenger exchange is 7.25 s. For exchanges involving up to five boarding or alighting passengers, the average duration increases to nearly 13 s, and for all measured stops, the average stop duration is 12.17 s. The stop "8 Maja" towards the city center was excluded from the calculations due to extended stop times caused by traffic lights.

5 Conclusions and Summary

Based on the measurements obtained, it can be inferred that:

- The possibility of skipping stops can reduce tram travel time on routes longer than one stop. However, it should be noted that this may not occur in every instance, as the likelihood of a vehicle stopping at an request stop can vary depending on the day of the week, time of day, or season.
- Most "request" stops in the Metropolis GZM tram network are justified based on a small sample size. For a more thorough analysis confirming the justification of stop statuses, it would be necessary to use supporting technologies or involve a larger number of individuals conducting manual measurements.
- In the Wrocław tram network, there are stops where implementing an request status is possible. One stop clearly indicates this possibility ("Karłowicza"), and for the next five stops ("Tarczyński Arena (Lotnicza)", "Glinianki", "Jeleniogórska", "Stadion Olimpijski", "8 Maja"), this status is worth considering.
- The presence of major event venues may suggest the need for special solutions such as changing the status of stops during their duration or disqualifying these stops from consideration.
- The average tram stop duration without passenger exchange was 7.25 s for the surveyed stops in Wrocław. Taking into account deceleration and acceleration times, skipping at least one stop could potentially reduce travel time by several seconds.

The implementation of request stops in the tram network could potentially reduce travel times by several seconds, even up to several minutes. However, this will only be possible if the designated stop status is justified and reflects actual passenger interest. For trips towards the city center, potential time savings can be estimated based on the probability of not stopping at specific stops, and then incorporated into the travel time calculations for the rest of the route segment. Similarly, for trips departing from the city center, potential time savings may allow for a reduction in average travel time on that segment, potentially reducing schedule delays.

Analysis of stop solutions, operational times on public transport lines, including passenger exchange times, but also research on the demand itself at specific stops should be the area of conducted analyses. Changes in the nature of the operation of stops can be, as part of the first steps, mapped and simulated in computer environments. An example of this type of activities for bus stops is research [23], in which operating bus stops were simulated and various scenarios of changes in the volume of passenger exchange and the impact of these changes on the operation of specific bus lines and globally the entire network were analyzed.

To proceed with the development of a model for validating the feasibility of introducing request tram stops, it is imperative to gather a more extensive dataset of measured tram journeys. This approach will bolster the model's effectiveness, particularly if supported by technological advancements. Enhanced data collection can be facilitated through passenger counting devices and real-time vehicle tracking systems. Collaboration with transportation authorities and operators is essential to conduct these studies effectively, contingent upon the availability of pertinent information.

References

1. Bauer, M., Richter, M.: Statistical analysis of time lost by trams before departure from stops. Civ. Eng. **12–14** (2006)
2. Molecki, B., Soczówka, A.: Przystanki na żadanie w transporcie kolejowym. TTS Tech. Transp. Szyn. **11–12**, 27–34 (2019)
3. Zych, M.: Tramwajowe przystanki na żądanie - przegląd rozwiązań. Logistyka. **4**, 3836–3846 (2014)
4. Molecki, A.: Metoda określania statusu przystanków w miejskim transporcie zbiorowym. Transp. Miej. i Reg. **12**, 28–30 (2004)
5. Zych, M.: System przystanków na żądanie – Budapeszt jako wzorzec. Transp. Miej. i Reg. **11**, 32–35 (2013)
6. Czerepicki, A., Krukowicz, T., Górka, A., Szustek, J.: Traffic light priority for trams in Warsaw as a tool for transport policy and reduction of energy consumption. Sustainability. **13**, 4180 (2021)
7. Górka, A., Czerepicki, A., Krukowicz, T.: The impact of priority in coordinated traffic lights on tram energy consumption. Energies. **17**, 520 (2024)
8. Furth, P.G., Muller, T.H.J.: Optimality conditions for public transport schedules with timepoint holding. Public Transp. **1**, 87–102 (2009)
9. Alonso, B., Moura, J.L., Dell'Olio, L., Ibeas, Á.: Bus stop location under different levels of network congestion and elastic demand. Transport. **26**, 141–148 (2011)
10. Jørgensen, F., Solvoll, G.: Welfare consequences of request stops at transport services with low demand. Eur. Transp. Res. Rev. **12** (2020)
11. Gomes, R., Pinho de Sousa, J., Galvão Dias, T.: Sustainable demand responsive transportation systems in a context of austerity: the case of a Portuguese city. Res. Transp. Econ. **51**, 94–103 (2015)
12. Warszawski Transport Publiczny: Zmiana charakteru przystanku METRO MŁOCINY 16. Warszawa. [Online]. https://www.wtp.waw.pl/zmiany/2023/02/23/zmiana-charakteru-przystanku-metro-mlociny-16/ (2023). Accessed 28 Jun 2024
13. Zarząd Transportu Metropolitarnego: ZTM wybór linii, autobus, tramwaj, trolejbus - rozkład jazdy. [Online]. https://rj.metropoliaztm.pl/ (2024). Accessed 04 Jul 2024

14. WPK: Archiwalne rozkłady jazdy. [Online]. http://rozklady.wpk.katowice.pl/archiwalne/. Accessed 28 Jun 2024
15. "Wayback Machine." [Online]. http://web.archive.org/. Accessed 28 Jun 2024
16. Tramwaje Śląskie, S.A.: Archiwum komunikatów – rok 2009. [Online]. https://www.tram-silesia.pl/www/index.php/2992/archiwum-komunikatow-rok-2009/ (2009). Accessed 04 Jul 2024
17. miastozabrze.pl: Przystanki tramwajowe "Biskupice Schronisko" zostaną włączone do obsługi pasażerskiej. [Online]. https://miastozabrze.pl/2024/02/20/przystanki-tramwajowe-biskupice-schronisko-zostana-wlaczone-do-obslugi-pasazerskiej/ (2024). Accessed 04 Jul 2024
18. MPK S.A. w Krakowie: Kieszonkowy rozkład linii 1. Kraków. [Online]. https://fotozajezdnia.pl/details.php?image_id=4055 (1992). Accessed 28 Jun 2024
19. MPK S.A. w Krakowie: Rozkład jazdy Miejskie Przedsiębiorstwo Komunikacyjne S.A. w Krakowie. [Online]. https://rozklady.mpk.krakow.pl/ (2024). Accessed 28 Jun 2024
20. Smajdor, M.: Oznaczenia przystanków. [Online]. http://www.kmk.krakow.pl/artykul_oznaczenia_przystanki.html (2024). Accessed 28 Jun 2024
21. Zarząd Transportu Publicznego w Krakowie: Komunikaty archiwalne - 2023. [Online]. https://ztp.krakow.pl/kmk/komunikaty-archiwalne/2023-pl (2024). Accessed 28 Jun 2024
22. ZDiTM Szczecin: Rozkład jazdy według linii. [Online]. https://www.zditm.szczecin.pl/pl/pasazer/rozklady-jazdy/wedlug-linii (2024). Accessed 28 Jun 2024
23. Fernandez, R., Tyler, N.: Effect of passenger-bus-traffic interactions on bus stop operations. Transp. Plan. Technol. **28**, 273–292 (2005)

Assessment of the Geometrical Condition of the Tram Track with the Use of Selected Survey Methods

Izabela Wilczyńska[1] (✉), Bartłomiej Ćmielewski[2], Agnieszka Tubis[3], and Aleksander Rau[4]

[1] Wroclaw University of Environmental and Life Sciences, Institute of Geodesy and Geoinformatics, Wroclaw, Poland
izabela.wilczynska@upwr.edu.pl

[2] Wroclaw University of Science and Technology, Faculty of Architecture / Laboratory of 3D Scanning and Modeling, Wroclaw, Poland
bartlomiej.cmielewski@pwr.edu.pl

[3] Wroclaw University of Science and Technology, Faculty of Mechanical Engineering, Wroclaw, Poland
agnieszka.tubis@pwr.edu.pl

[4] GeoSter Company, Wroclaw, Poland
biuro@geoster.pl

Abstract. Evaluation of the geometric condition of the tram track is performed according to train instruction Id-14 “Instruction for measuring, testing and assessing the condition of the tracks” developed by PKP PLK S.A.. Hence, it corresponds to the requirements for railroad tracks. The experiment and measurement works were carried out in cooperation with MPK Wrocław (Urban Transport Company) on the section of the tram track at Mickiewicza Street in Wroclaw, Poland. This part of the track was excluded from the planned traffic due to bad conditions. The following survey techniques were used to compare and assess the geometric state: the LTrack electronic track gauge, tachymetric with a specialised adapter with a prism and terrestrial laser scanning. In addition, the geometric condition of the track was evaluated using the J and W5 indicators.

Keywords: Tram track inventory · Leica LTrack · Tachymetry · Terrestrial laser scanning

1 Introduction

The railway surface is a crucial element of the railroad, and in the light of increasing the permissible speed of rail vehicles and increasing the pressure on the track, it is required to increase its strength and safety. As a classic pavement structure, the track works in the elastoplastic area, accumulating permanent deformations of the ballast, which leads to the differentiation of elastic properties and damping along the track length during operation [1]. Therefore, it is vital to monitor the geometric condition of the track to

G. Sierpiński et al. (Eds.): TSTP 2024, LNNS 1257, pp. 145–158, 2024.
https://doi.org/10.1007/978-3-031-81777-9_10

more efficiently control track damage and optimise the schedule for repair and maintenance of Wrocław tracks [2]. An inventory of the gauges is also essential so that special care is taken in hazardous locations. The limitations of the indicator include, among others: the structure of tunnels and bridges, platform edges, platform shed roofs, fences and noise barriers, buildings (continuous type limitations) and semaphores, warning and manoeuvring discs, indicators, traction poles, footbridge supports and fi bridge lanterns, switch lights etc. (point type restrictions) [3, 4]. In geodesy, two measurement tendencies develop in parallel—one concerns collecting as much information as possible during one trip, the other aims at the shortest possible travel times [5, 6]. Significant in this aspect is the possibility of using GNSS satellite techniques to determine the position [7].

Although the technical requirements for railways and urban transport differ significantly, these issues also apply to tram tracks. Currently, the tracks are built under the Road Act [8], while the inventory is carried out according to the rules applicable in the railways [9]. Contrary to railroads, tram routes do not have an organised set of regulations corresponding to railway guidelines and instructions [10]. Improving the condition of the tracks, and thus the comfort of travelling by public transport, and MPK Wrocław's search for the most advantageous method of track inventory is the main reason for conducting the research work.

2 Guidelines for the Assessment of the Condition of Railways

2.1 Instruments and Tools for Measuring

The measurements devices should meet the requirements according to the Id-14 instruction. Then, with the help of a track gauge, vertical and horizontal parameters are determined. The vertical parameters are: the cant—this is the size of the difference in the height of rail tracks in curves, the twist is the ratio of cant measured on two adjacent sections to the distance between them, the trough is the unevenness of a single bar track in the vertical plane, i.e. deviation the rail running surface from the reference level defined by the chord joining the wheel contact points of the measuring vehicle. In the case of horizontal parameters, the following are distinguished: track width—this is the distance between two rails measured 14 mm below the top surface of the rail. Track width gradient—this parameter determines the difference between the spacing of rails within the measurement base with a length of 1 m. Horizontal railroad unevenness—parameter this is defined similarly to the determination of the value of the hole. Finally, the horizontal deviation of the track within the 10-m base, which is a chord joining the two extreme measurement points, is determined [9].

2.2 Parameters of Railroad Condition Assessment

The results obtained in the fieldwork were developed and analysed following the PKP instructions for measuring, testing and assessing the condition of the tracks. Under §7 on the statistical evaluation of the rail condition, the results were analysed using two approaches. The first was the determination of the synthetic track condition index—J. The second method was the determination of the five-parameter defectiveness. The J index is an objective quantity describing the condition of the track. This value is independent of the speed at which rail vehicles travel along this section. The following parameters are necessary to determine the J index: —vertical irregularities, —horizontal irregularities, —track twist, —track gauge. After determining the given values, the J index can be calculated using the formula (formula 1):

$$J = \left(\left(S_z + S_y + S_w + S_e / 2\right)\right) / 3{,}5 \tag{1}$$

where:

S_z—the value of the standard deviation of vertical irregularities,
S_y—the standard deviation of the horizontal anomalies,
S_w—the standard deviation of rail twist,
S_e—the value of the standard deviation of the track width.

After receiving the value of the J index, it is possible to determine the condition of a given section of the track. First, the ID 14 instruction specifies the maximum values of the J parameter in various spweed ranges of rail vehicles moving along a given section. For example, for railways, speeds of up to 200 km/h are planned, but for tram tracks, the maximum speed of 50 km/h has been adopted, for which the maximum value of the J index is 8.2. The second method of determining the track condition is to determine the five-parameter W5 defect, i.e. the value that changes with the change of the speed of the rail vehicle on a given track section. To resolve this amount, the following formula (formula 2) should be used:

$$W_5 = 1 - \left[\left(1 - W_e\right)^* \left(1 - W_g\right)^* \left(1 - W_w\right)^* \left(1 - W_z\right)^* \left(1 - W_y\right)\right] \tag{2}$$

where:

W_e—the size of the track gauge defect,
W_g—the amount of track cant defect,
W_w—the amount of track twist defect,
W_z, W_y—arithmetic means of defectiveness of vertical and horizontal irregularities.

After receiving the parameter value W_5, the measured distance should be assigned to the four categories specified in Annex 5 to the Id-14 instruction.

3 Research Facility and Surveys

3.1 Measurement Network

The object on which the research works were carried out was the section of the tram track located at Mickiewicza Street in Wrocław. It is an episode excluded from regular traffic. It is used only when necessary to designate detours for trams, during breakdowns or other unforeseen accidents on nearby tram routes. Therefore, a section of about 800 m was selected for measurements, located between the intersection with Wystawowa Street and the junction with Mikołaja Kopernika Street. In the central part of the measuring section, there are two asphalt crossings through the track, which caused minor problems when measuring with the track gauge. Therefore, the first stage of measurements was the establishment of the measurement network. The network was designed in compliance with the requirements for geodetic networks of engineering objects [11]. It consisted of 18 stabilized points within the track, 6 points on the opposite side of the street and rangefinder plates located on the path's poles. For stabilization, nails were used in the area of the track and rangefinders, which were glued reflective signals, installed on poles located several dozen centimeters from the track, evenly spaced along the entire measuring section.

The measurement of the control points was made at the turn of 2019/2020 using the Trimble s8 robotic electronic total station, capable of angular accuracy of 0.5″. The next step was to measure the precise levelling of all points of the network. This measurement was performed using the Leica DNA 03 code leveler and dedicated code bars with supports.

3.2 Measurement with a Track Gauge

The measurements with a track gauge should be carried out in intervals of a maximum of 10 m on straight sections and every 5 m on curves [9]. The Leica LTrack system was used to perform the measurements. The set includes a trolley of a track gauge equipped with a precise tiltmeter, a mirror (Fig. 1), and a Leica TS16—robotic total station. The measurement points were marked on the sides of the rails with white paint, which allowed for subsequent measurements in the same places.

Of course, to use this kind of system for the inventory of tram tracks, it must be previously adapted to the conditions prevailing in the city. First of all, in the city, some rails are integrated with the road surface, and in particular, grooved or block rails are used at turnouts or curves [12].

Measurements with the track gauge were highly smooth, mainly due to the complete L-Track system, the track gauge itself and a dedicated Leica TS16 robotic total station with a mirror and comprehensive computer software for data processing and visualisation. The data had to be exported in a suitable format that the Rail Office software could read to process the data further. The final result of the data preparation was a table containing XYH coordinates of given points on cross-sections. Each section consists of two points, one on each rail track [13].

Fig. 1. Leica track gauge when taking track measurements.

Fig. 2. Trimble s8 while taking track measurements.

3.3 Total Station Measurement

The next stage of the measurement work was the measurement of track parameters using the tachymetric method. For this purpose, a Trimble S8 robotic total station and a mirror with an adapter adapted to measure rails were used (Fig. 2). This total station has an angular accuracy of 0.5″, making it one of the most accurate instruments of this type. Importantly, the 0.5″ accuracy is also maintained by using the Autolock function, which allows for automatic Trimble mirror detection. Distance measurement using a mirror takes place with an accuracy of 1 mm + 1 ppm, and the maximum range with a classic prism is 3000 m [12]. Thanks to a mirror equipped with a transmitter and a robotic total station, the entire measurement could be performed by one person, which is an undoubted advantage of this measurement method in terms of economy. Furthermore, the mirror was set at the same points where the track meter was previously stopped during the measurement. Therefore the results of the combination of both processes could be compared.

The final result of the tachometric measurement was a set of coordinates of all measured points in the following sections. The first column contained the numbers of consecutive measurements, then the XYZ coordinates of each point and the information to which rail track the end belongs—left track—sz_l, and right track—sz_p. Finally, while further comparing the measurement results, the unified nomenclature specifies the left-hand path as step 1 and the right-hand path as step 2. The data was processed in Microstation programs and the proprietary program of the Geoster company.

3.4 Measurement with a Terrestrial Laser Scanner

The last of the methods used were terrestrial laser scanning. The time to perform the measurements on one station was about 90 s because there was no need to manually set the scanner on a tripod at each station. The scanning was performed using the Leica P40 scanner (Fig. 3), ensuring the desired measurement accuracy while maintaining a short measurement time—the scanning process was about 58 s. The Leica P40 scanner is characterised by high accuracy and low measurement noise.

Fig. 3. The Leica P40 scanner while working on a developed track trolley.

The Leica ScanStation system allows obtaining the most elevated quality 3D data and HDR imaging while maintaining a high scanning speed of 1 million points in 1 s. Such a high-quality scan can be made up to a distance of 270 m from the scanning station. The unparalleled advantages of the Leica P40 scanner include high distance measurement accuracy of 1.2 mm + 10 ppm over the entire range, horizontal and vertical angular accuracy of 8″, and the accuracy of the three-dimensional position determination is 3 mm with a target length of up to 50 m. However, during the measurements, it did not even get close to the maximum value of 50 m [14], due to the small matter of the angle α when the laser beam hits the rail surface. This was due to the height at which the scanner was mounted (about 30 cm above the railhead level), which would result in very low accuracy in the location of the points with longer sights.

4 Measurement Results

The measurement results were analyzed analytically in accordance with the guidelines used in the railway industry. The results of the analysis of this part of the work are presented below.

4.1 Analysis of Measurement Results—Index J

The results obtained as the fieldwork were developed and analysed following the PKP instructions for measuring, testing and assessing the condition of the tracks. Following §7 on the statistical evaluation of the rail condition, the results were analysed using two approaches. The first was the determination of the synthetic track condition indicator—J. The second method was to determine the five-parameter defectiveness. The J index is an objective quantity describing the condition of the track. This value is independent of the speed at which rail vehicles travel along this section. Therefore, the standard deviations of the following parameters should be calculated in advance: —vertical irregularities, —horizontal irregularities, —track twist, —track gauge, to determine the J index. For this purpose, according to the instructions of PKP Id-14 [9], the formula should be used for calculate standard deviation.

4.2 Measurement Results Analysis—Index W5

To assess the defectiveness, it is necessary to obtain the five-parameter defectiveness index, for each of them, the ratio of the total length of the section on which the values of deviations considered acceptable have been exceeded should be calculated. Therefore, we determine the faults of the parameters from the following formula (formula 3):

$$W = n_p / n \tag{3}$$

Table 1. Summary of the W5 index values in individual classes.

Heading level	Class C	Class B	Class A
Track gauge	0.473	0.645	0.864
Scanner	0.451	0.644	0.870
Total station	0.462	0.630	0.850

where:

n_p—the number of acquired signal samples, the value exceeds the maximum deviations on the tested section,
n—number of all signal samples on a given segment.

The research work's calculations of track condition indicators, using the above formulas and guidelines. The first stage of the chamber works was to organize the data. Finally, signal samples in which the value exceeded the allowable value were listed. For all parameters, the values are classified as classes A and B. Calculations was performed for all parameters (formula 2) that is: W_e—the size of the track gauge defect, W_g—the amount of track cant defect, W_w—the amount of track twist defect, W_z, W_y—arithmetic means of defectiveness of vertical and horizontal irregularities.

4.3 Summary of the Measurement Results

The results of all measurement methods were very similar, which is desired effect in terms of the correctness of each one. We assess the geometrical condition based on the W5 index, while the synthetic track condition indicator "J" is an objective assessment of the track condition, independent of the permissible line speed. It is based on the standard deviations of vertical and horizontal irregularities, track twists and track gauges. In all methods, it is convergent and amounts to 5.14, 5.70 and 5.53, respectively. Detailed results of the assessment with the W5 index are presented in Table 1.

After comparing the results, it is visible that they are convergent. Furthermore, identical results were obtained for all measurement methods when calculating the W5 index while maintaining writing precision to decimal parts. Thus, the W5 index was within limits for each technique, determining the track as good enough geometry only for class C. For Class B, a W5 value of 0.6 was the cut-off between acceptable track conditions and unserviceable.

5 Assessment of the Condition of the Tram Track

5.1 Development of the Concept for the Track Condition Assessment

In the analysis of the dynamics of rail vehicles, vehicle traffic is considered divided into two components. The first component is movement, and the second is movement disorders. In this context, studies have been conducted to define a set of track quality indicators to test its condition [1, 15]. Another way to assess the track's state may be the analysis of the subgrade and classification according to its condition. The concept of the system for determining the technical condition of tram crossings was drawn up in Gdańsk, analysing the material from which they are made. Apart from crossings for pedestrians and cyclists, there are 223 crossings with a total length of 2942.97 m on the Gdańsk tram network [16].

When developing the concept of track assessment, it is worth considering solutions used in crane tracks, the parameters of which are strictly defined by standards. The safe use of cranes and their failure-free operation depend on meeting all its components and crane tracks [17]. Improper track arrangement and their subsequent condition resulting from the operation of the crane increases the running resistance, which has an impact on faster wear of the road wheels. In an extreme case, it may cause the crane to fall out of the tracks [18, 19], which takes place in urban rail transport.

5.2 Assessment of the Condition of the Tram Track in Wrocław

In 2011, a detailed inventory of Wrocław tracks was carried out. At that time, the J index's limit values were adopted to assess the track condition: very good condition 0.0–2.9, good condition 3.0–6.9, average condition 7.0–10.9, bad condition 11.0–14.9, very bad condition from 15.0 [20]. The ID-14 instruction does not provide for such a classification, but only the limit values for the J index for individual speeds. Nevertheless, as can be seen in the sketch from the end of 2015, many track sections were described as very good condition. Some of them were newly built tracks.

There is no tram line in Wrocław, which was characterised by a minimum good condition of the track along its entire course. Only in 2018, according to data collected by the Municipal Transport Company in Wrocław, 79 derailments were recorded. In the following years, 2019 and 2020, this number exceeded 100 derailments. Many factors influence it, but the analysis results by Deutzer, whose services were also used by e.g. Paris, Sydney, Vienna and Brussels, show that sections with sufficient or unsatisfactory geometric conditions constitute 78% of the entire network [21].

In 2018, the Roads and City Maintenance Authority (ZDiUM) in Wrocław carried out an audit of the condition of the tracks. It shows that 58% of the track is in very good condition, 31% in average condition and only 11% in very bad condition [22]. After performing specialised measurements in 2020 and assessing the track according to the Id-14 instruction, the indicators suitable for railways are as follows: 5.43% in very good condition, 5.92% in good condition, 54.68% in average condition and 33.97% in bad condition [23] (Table 2).

The change in the track condition within 2 years is significant, not only in terms of use but also in terms of the assessment method and classification criteria. In 2020, an

Table 2. Comparison of the track condition in 2018–2020.

State	2018	2020	2020 (aggregation)	Difference
Very good	58%	5.43%	11%	−47%
Good	–	5.92%	–	
Average	31%	54.68%	55%	+24%
Bad	11%	33.97%	34%	+23%

assessment of the track condition was adopted based on the W5 index used in railways. Many sections are in very good and good condition according to the evaluation from 2018, currently in the assessment of railway indicators are described as a poor or average condition.

5.3 Causes of the Detentions in 2019 r.

The state of the tram track should undoubtedly be performed based on standards separate from those used in railways. I suggest starting the concept of the track condition analysis with the classification of unplanned tram stops and their location. Currently, this list is not kept harmonised. In 2019, 1084 events took place, described in about 100 ways. The event's location is indicated in various ways—most often only with the street's name. We grouped the data from 2019 into 19 main reasons (Fig. 4). One hundred nine stops were due to damage to the rail. Thus, the assessment of the condition of the railhead is essential in 10% of unplanned stops. It is a significant number of stops among the reasons we have real influence. Factors that we cannot eliminate or that would be very difficult include: blocked passage, accidents, lack of voltage, health problem of a passenger or driver, road conditions (e.g. branches on the net, slippery rails), damage to the network. They account for 39% of detentions. 42% of the events are related to the damage or failure of the tram. It should be emphasised that the track condition translates into the wear of some tram components, e.g. rolling wheels.

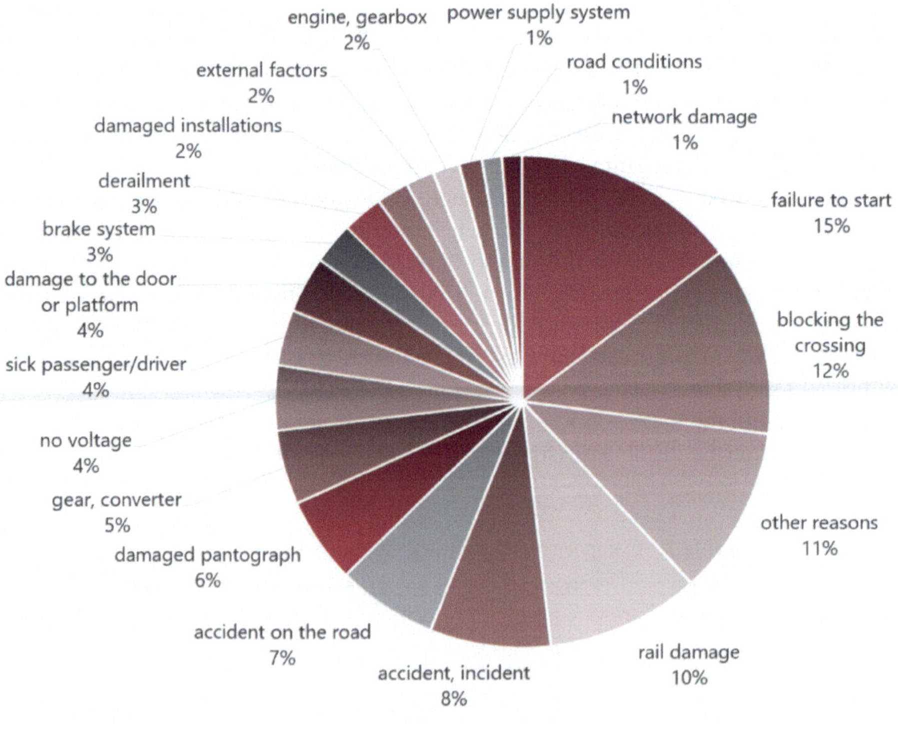

Fig. 4. Total causes of the detentions in 2019.

The replacement of the tram rolling stock in Wrocław is constantly modernised. MPK Wrocław (Urban Transport Company) has 285 trams, and the low floor is available in 67% of the tram fleet [24].

To make it easier to complete the database, a system should be created that would allow drivers to easily report an incident—e.g. in the form of an application, providing a location on the map (suggested for approval or with the possibility of indicating), the date and time of the start of the event, names of stops) The reason from the selectable list is on two levels of detail. The application should be able to provide DIP (Dynamic Passenger Information) and its own additional comments.

5.4 Spatial Analysis of Unplanned Stops

Spatial analyses are widely used in the world also in terms of road safety [25–27]. A map was made with their places of occurrence to determine the correlation between the events and their location. As I mentioned earlier, the description of the incident is often limited to the street name only, so these will be diagrams on the map accumulated on a given street. Unplanned stops from 2019 were taken into account, while the condition of the track, classified into four groups, was presented on the map based on the results of measurements carried out in the first half of 2020.

The determination and classification of the reasons for the stops will be the basis for developing the concept for assessing the track condition. As can be seen on the cluster map (Fig. 5), the bad condition of the track does not always translate into the number

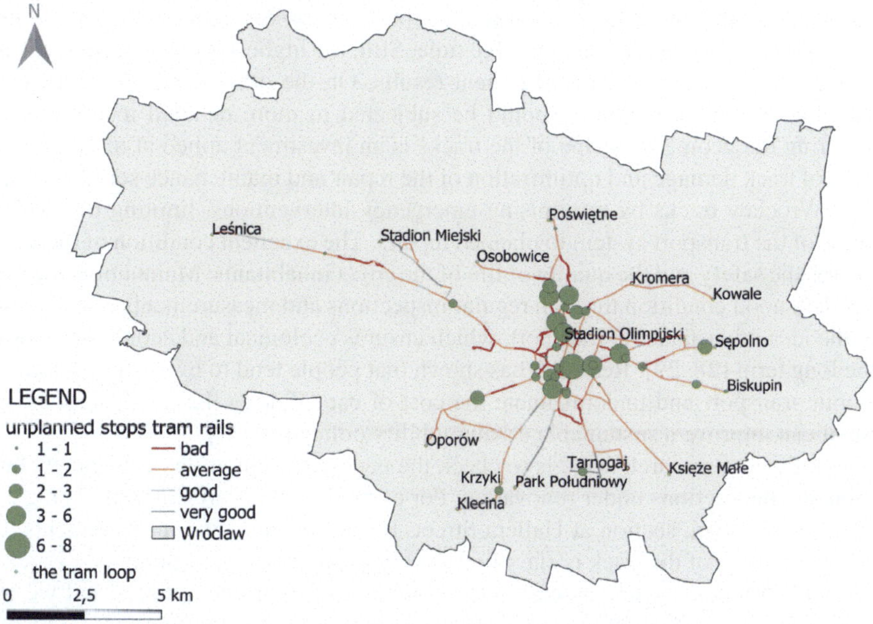

Fig. 5. Map of clusters of unplanned stops related to the poor condition of rails against the background of tram track layout.

of unplanned stops resulting from this condition. The most significant unplanned stops occurred at Dominikański Square, where the track condition was considered average. Therefore, this map is limited to 109 bus-related events that the log keepers have selected.

6 Summary and Conclusions

Many indicators make it possible to determine the geometric condition of railway tracks, classify them. However, no clear guidelines specify the methods of measuring the tram track and assessing its condition. The track condition is evaluated based on the ID-14 instruction used in railways. These requirements are not adequate to the requirements for a tramway track. Therefore, work is underway to adapt the instructions for the control of the track geometric condition to the needs of MPK.

Tests were carried out on the section excluded from the planned traffic at Mickiewicza Street to check the effectiveness of various measurement methods. Field measurements were made from January to May 2020, translating into their results due to temperature changes and track wear. Furthermore, the position of the compared cross-sections was marked with paint, which could translate into coordinate differences. Despite using different methods, the obtained results were identical for the W5 index and similar results for the J index. It is worth noting that the test section did not include arcs that could show differences in the condition assessment between individual measurement methods.

Considering the multifaceted nature of the problem, assessing the tram track condition should occur in the shortest possible time. Still, the highest level of accuracy is not necessary, as shown by the measurement results. On the other hand, the assessment revealed poor track conditions should be subjected to more detailed measurements. Monitoring the geometric shape of the tracks is an investment aimed at more efficient control of track damage and optimisation of the repair and maintenance services schedule for Wrocław tracks by minimising emergency interventions, limiting the ineffectiveness of the transport system to planned repairs. The excellent condition of the tracks increases the safety and the quality of life of the city's inhabitants. Maintaining the tram network in good condition through regular inspections and measurements is also in line with the idea of sustainable transport, which ensures ecological and economic stability in the long term [28, 29]. Research has shown that people tend to overestimate the cost of public transport and underestimate the cost of cars. Taking these calculations into account can improve a sustainable urban mobility policy [30, 31].

The following research stage is to check the geometric condition of the newly built section and the sections under renovation. For example, works are currently being carried out on the track section at Hallera Street. If, as a result of the measurements, the specific condition of the track is classified as very good, it will be a guideline to stick to the J and W5 indexes when assessing the condition of the railway track, and we will limit the changes to the procedures to defining new measurement methods. However, with high probability, the J and W5 indicators will have to be modified to adapt to the conditions imposed on tram tracks.

References

1. Kostrzewski, M., Konowrocki, R.: Estimation of selected sectors of test track using quality track indicators. Pojazdy Szynowe. **1**(2), 1–17 (2014). https://doi.org/10.13140/RG.2.1.2340.1043
2. Chiacchiari, L., Loprencipe, G.: Measurement methods and analysis tools for rail irregularities: a case study for urban tram track. J. Mod. Transp. **23**(2), 137–147 (2015). https://doi.org/10.1007/s40534-015-0070-6
3. BN-89-9396-05/03 Norma branżowa, komunikacja miejska, Skrajnia budowli, Tabor tramwajowy (Industry standard, public transport, Building gauge, Tram rolling stock) (1990)
4. Poliński, J.: Pomiary ograniczeń skrajni budowli na kolejach niemieckich. Railway Report, Zeszyt 189, pp. 45–56 (2020). https://doi.org/10.36137/1895P
5. Strach, M.: Modern surveying techniques in the process of modernization and diagnostics of railway track geometry. Wydawnictwo AGH, Kraków (2013)
6. Strach, M., Piekarz, M.: Modern equipment used in rail-track measurements. Problemy kolejnictwa. **148**, 48–60 (2009)
7. Strach, M.: Ocena możliwości wykorzystania techniki satelitarnej RTK GPS do regulacji osi torów kolejowych. rozprawa doktorska Kraków (2003)
8. Ustawa z dnia 10 kwietnia 2003 r. o szczególnych zasadach przygotowania i realizacji inwestycji w zakresie dróg publicznych (Act on specific principles for the preparation and implementation of investments in public roads)
9. PKP Polskie Linie Kolejowe S.A.: Instrukcja o dokonywaniu pomiarów, badan i oceny stanu torów Id-14 (D-75) (Instructions for making measurements, testing and assessment condition of the Track Id-14). PKP Polskie Linie Kolejowe S.A., Warszawa (2005)
10. Grabias, P.: Selected measurement techniques in the study on the technical condition of tram tracks and loops. Transport miejski i regionalny, 19–23 (2019)
11. PKP Polskie Linie Kolejowe S.A.: Standard techniczny określający zasady i dokładności pomiarów geodezyjnych dla zakładania wielofunkcyjnych znaków regulacji osi toru Ig-7 (Technical standard specifying the principles and accuracy of geodetic measurements for the installation of multifunctional track axis regulation signs Ig-7) PKP Polskie Linie Kolejowe S.A., Warszawa (2018)
12. Trimble GmbH: Datasheet - total station Trimble s8. Geotradingpolska.pl. https://totalstations.co/wp-content/uploads/2017/07/Trimble-S8-Edited-1.pdf (2021)
13. Szołdra, M.: Functional analysis of selected measurement methods used for the inventory of the tram track. Master Thesis, Wrocław (2020)
14. Leica Geosystems AG part of Hexagon: Leica ScanStation P30/40 – Technical Specifications. https://w3.leica-geosystems.com/downloads123/hds/hds/general/brochures-datasheet/Leica_ScanStation_P30-P40_Plant_DS_en.pdf (2021)
15. Chudzikiewicz, A., Droździel, J., Sowiński, B.: Practical solution of rail vehicle and track dynamics monitoring system. Struct. Health Monit. II Book Ser. Key Eng. Mater. **518**, 271–280 (2012)
16. Grulkowski, S., Zariczny, J.: Concept of evaluation of technical state of tramway crossings. Zeszyty Naukowo-Techniczne Stowarzyszenia Inżynierów i Techników Komunikacji w Krakowie. Seria: Materiały Konferencyjne. **2**, 109–119 (2013)
17. Bryś, H., Ćmielewski, K., Kuchmister, J., Gołuch, P.: Verification measurements of crane rail track geometry. Przegląd Geodezyjny. **90**(9), 16–22 (2018). https://doi.org/10.15199/50.2018.9.1
18. Anigacz, W.: Modyfikacja geodezyjnych metod projektowania rektyfikacji jezdni podsuwnicowych. Wydawnictwo Wyższej Szkoły Inżynierskiej w Opolu, Monografia (1992)

19. Anigacz, W.: Weryfikacja parametrów geometrycznych suwnic w czasie montażu i eksploatacji. Transport Przemysłowy. **2**(4), 38–40 (2001)
20. Makuch, J.: Tram track geometry measurement by digital track gauge in Wrocław. Przegląd Komunikacyjny. **8** (2014)
21. Matejuk, T.: 11% torowisk tramwajowych we Wrocławiu jest w dobrym stanie [RAPORT]. Nasze Miasto 04.01.2021. https://wroclife.pl/nasze-miasto/torowiska-tramwajowe-wroclaw-raport (2021)
22. Matejuk, T.: NIK criticises public transport in Wrocław [REPORT]. Nasze Miasto 07.07.2020. https://wroclife.pl/nasze-miasto/nik-komunikacja-miejska-wroclaw/ (2021)
23. Deutzer Technische Kohle GmbH: Measurement Report 618b/2020. https://www.akcjamiasto.org/wp-content/uploads/2021/01/Raport_2020_Pomiar-toru-i-szyn_Wroclaw_618b_PL.pdf (2021)
24. MPK Wrocław: Nasze pojazdy. https://mpk.wroc.pl/o-mpk/nasze-pojazdy (2024)
25. Dereli, M.A., Erdogan, S.: A new model for determining the traffic accident black spots using GIS-aided spatial statistical methods. Elsevier Transp. Res. Part A Policy Pract. **103**, 106–117 (2017). https://doi.org/10.1016/j.tra.2017.05.031
26. Young, J., Park, P.Y.: Hotzone identification with GIS-based post-network screening analysis. J. Transp. Geogr. **34**, 106–120 (2014). https://doi.org/10.1016/j.jtrangeo.2013.11.007
27. Park, H.-C., Yang, S., Park, P.Y., Kim, D.-K.: Multiple membership multilevel model to estimate intersection crashes. Accident Anal. Prevent. **144** (2020). https://doi.org/10.1016/j.aap.2020.105589
28. Szymalski, W.: Bringing European transport policies on the climate neutrality track: assessment of and recommendation on members states and EU transport policies. CAN-Europe. https://www.pine.org.pl/project/publikacje/ (2021)
29. Szymalski, W., Kassenberg, A., Stainforth, T.: Implementing the European Green Deal at local level: lessons on the role of cities in Poland in planning climate policy. Mitigation, Adaptation, and Just Transition. Think2030 policy paper by the Institute for Sustainable Development and the Institute for European Environmental Policy (2021)
30. Meira, L.H., de Mello, C.A., Castro, Y.M., Oliveira, L.K., Nascimento, C.O.L.: Measuring social effective speed to improve sustainable mobility policies in developing countries. Transp. Res. Part D Transp. Environ. **78** (2020). https://doi.org/10.1016/j.trd.2019.12.002
31. Dubé, J., Legros, D., Devaux, N.: From bus to tramway: is there an economic impact of substituting a rapid mass transit system? An empirical investigation accounting for anticipation effect. Transp. Res. Part A Policy Pract. **110**, 73–87 (2018). https://doi.org/10.1016/j.tra.2018.02.007

Fare Integration Under Conditions of Suburbanisation and Location of Transfer Centres Outside Urban Areas

Grzegorz Dydkowski[1], Wiktoria Gałecka[1], and Anna Urbanek[1] (✉)

[1] Department of Transport, University of Economics in Katowice, Katowice, Poland
grzegorz.dydkowski@ue.katowice.pl;
wiktoria.galecka@ue.katowice.pl; anna.urbanek@ue.katowice.pl

Abstract. The paper presents processes of suburbanisation and location of selected elements of the passenger transport infrastructure in suburban areas as the factors, which should cause intensification of implementing solutions from the field of tariff integration of the urban and regional passenger transport. In suburban areas, due to a low density of population and frequently scattered residential development, it is difficult to achieve a high quality of public transport services. In addition, the suburban and regional transport can differ from the urban transport in tariff systems and tariff regulations. The tariff integration, also with the use of IT, will allow to improve the competitiveness of the public transport with movements by cars.

Keywords: Fare integration · Suburbanization · Public transport · Transfers · Urban transport

1 Introduction

The suburbanisation, location of residential housing, but also of other functions in suburban areas, is one of processes, which occur in various urban centres and in their surroundings worldwide. This is a multidimensional and complex process with its origins and effects for the spatial management, resources management, the economic, social, and environmental domains, functioning of cities and suburban areas, and also for transport. There is a great dependence of spatial development of cities on means of transport, and in the case of suburbanisation, in particular on the availability of cars and popularisation of individual motorisation [1]. The processes of suburbanisation grew stronger earliest in the United States and Canada [2], after World War II the development of housing and of low-rise buildings intensified in the neighbourhood of big cities, in the 1970s in Western European states [3], and at the turn of the twentieth and twenty-first century also in Poland [4–11].

A few sources of suburbanisation may be indicated, from the point of view of people making choice these are lower prices, and also easier availability of land, better living conditions, like the neighbourhood of green and recreational areas and thereby no

G. Sierpiński et al. (Eds.): TSTP 2024, LNNS 1257, pp. 159–177, 2024.
https://doi.org/10.1007/978-3-031-81777-9_11

arduous neighbourhood of say, areas of industrial or transport functions, and also arduousness resulting from dense development and high population. The residence in urban areas means also cleaner air. However, the condition consists in a possibility of getting quickly to the city, which most frequently is by car, and hence the availability and relative cheapness of this form of moving is the condition.

Suburbanisation processes are assessed in various ways, frequently environmental issues are emphasised, related to the generation of high car traffic and thereby increased demand for the road transport infrastructure [12–14], opinions are expressed about uncontrolled spatial development increasing the costs of municipal infrastructure, depopulation of city centres, creation of social divisions and neighbourhoods inhabited by low- and high-income people. The issues of time losses on commuting, difficult access for kids and youth to kindergartens, schools, or other places important for education, but also for development of interests, are raised [15–18]. However, suburbanisation is also related to benefits, apart from those mentioned before, this is also a strong possibility of choosing places of work, perhaps even stronger than for persons living in the city.

Together with changes in the functional and spatial structure of cities and the growing urbanisation, changes should also proceed in transport systems, including the passenger transport systems. From the point of view of organisation and management of public transport systems, it is necessary to draw attention to the fact that urbanisation processes may consist in development of areas, within administrative boundaries of a given unit, or via migration of residents and management of suburban areas. If in the former case, the development of areas more and more distant from city centres is related, on the level of specific unit, to parallel and integrated spatial planning and to development of urban collective transport planning, then in the latter case movements from suburban areas, but from another administrative unit, fall into the category of intercity movements, and in this case the issues of public transport are not always subject to relevant planning, and the very movements by public transport from suburban areas are not necessarily within solutions integrated with the urban transport. Integrated solutions in the urban public transport are frequently confined within administrative boundaries of cities or conurbations, and the integration with suburban areas is a secondary task.

Also airports are situated in suburban areas, outside urban areas, which become increasingly important in serving the conurbation needs; in addition, quite often new railway stations and stops are constructed outside cities, close just to airports, motorway junctions or ring roads, or in places, which are connected by the existing lines of suburban or conurbation railways. Also this, next to migration and settlement of people in suburban areas, is an important factor for implementation of solutions from the field of urban public transport integration with supra-local and regional transport systems.

2 Objective and Research Methods

A lot of research has been carried out and many publications are available in the field of organization and management of urban transport, in particular urban public transport. However, when designing solutions, administrative boundaries are often adopted, without taking into account the progressive suburbanization and, over time, the incompatibility of the existing city boundaries with the actual urban layout and the nature of the buildings, as well as economic and social ties, including the spatial distribution of daily commutes to work and study.

In addition to other activities leading to the integration of urban, suburban and regional transport systems, such as the coordination of timetables, the creation of interchange stations and common passenger information, the increasing of public transport attractiveness, and thereby meeting expectations of city residents and persons interested in using the public transport, requires to implement solutions, in which a given ticket or another transport document may be used not only in the public transport on suburban and regional lines carrying to a specific city centre, but also in the system of urban public transport. It requires taking actions aimed at introduction of fare changes, which unify part of fare provisions in the case of different public transport organisers/carriers, and expanding the scope of validity of existing ticket types or introducing new tickets, with a wider scope of entitlements.

The paper is aimed at presentation, under conditions of suburbanisation processes and location of selected elements of passenger transport infrastructure in suburban area as well as public transport organisation, of solutions and directions of fare integration for urban public transport systems with suburban areas lines.

Methods for analysing the existing solutions in the urban passenger transport and in regional transport were used at the paper writing, as well as methods for generalising, modelling, and synthesising. In particular the solutions in the public transport were analysed in terms of competence and tasks division between entities of the urban public transport—public authorities, organisers, and entities carrying out transport, competences in the field of fares establishment and of applied fare solutions. At the modelling and synthesising a rule was adopted, that solutions should be shaped in such a way, that on the one hand not to deteriorate the effectiveness of services provision and entities independence, and on the other hand to ensure fare integration facilitating the use of various public transport systems.

3 The Impact of Suburbanisation Processes on the Spatial Scope of Transport Fare Integration

There are numerous concepts of cities development and city types [19], also processes of their transformations and development over years proceed in various ways. Cities are diverse—in their economic, social, demographic, administrative, political structure, but also in their atmospheres, in their daily routines, in their sound volumes or smells; and, of course, also in their physical appearance, i.e. the built and non-built urban landscape [19]. The number of residents, the social-economic and functional-spatial structure changes, social and demographic processes are going on, and the expectations related to the living conditions in the city change. From the transport point of view, the basic features of urban space that should be strived for are balancing the sources and goals of traffic and intensifying spatial planning. The concentration enables to intensify contacts as well as the social, economic, and cultural exchange and by that—the origination of the added value and civilisation development. Economic benefits of greater concentration consist in increased productivity, functionality, and attractiveness, related to lower infrastructure costs and reduced transport needs [20]. The twentieth century experience has shown that the population density and intensity of development support productivity, functionality, and the value of city space. The concentration of development brings obvious economic benefits, such as lower infrastructure costs, lower transport costs, and economic justification of the urban transport development as well as time savings at movements [20].

Modern urbanisation processes, comprising vast areas beyond city reach, frequently feature parallel deconcentration and reconcentration of population as well as economic and social activity, leading to the origination of multi-centric metropolitan regions [21]. This results inter alia in growing domination of the transport and logistic system in the spatial and social-economic structure, combined with incommensurate growth of infrastructure (motorways, roads, ring roads, service stations, airport and its facilities, internal transport systems), which nodes gradually take away the functions from the old centre [22, 23].

The demographic development in Poland of cities surroundings throughout the post-war period is presented in the Fig. 1. The analysis shows that until around 1970 there was a systematic increase of the urban surroundings around all centers that were now at least subregional, and until around 1988—at least subregional. Since around 1988, there has been a marked increase in the "big five", which seems to be accelerating in the last inter-census decade. In turn, the growth rate of suburban zones around supra-regional centers (however, with a significant acceleration in 2011–2021), regional and sub-regional centers is clearly slowing down (in these two cases, the increase in the last decade is not so significant). However, the suburban zones of the smallest local

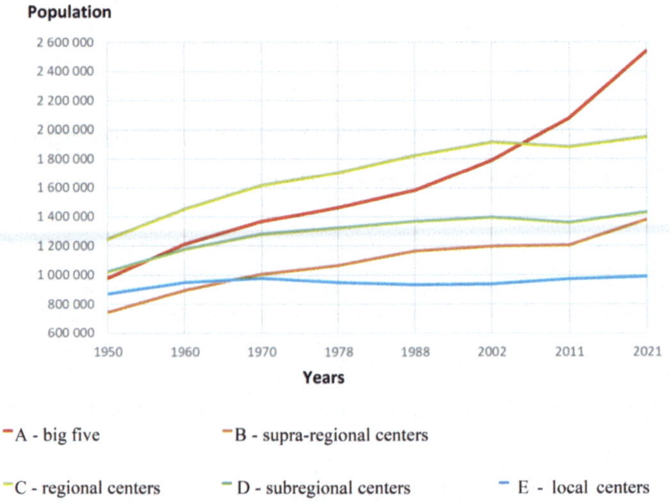

Fig. 1. Population changes in suburban zones in Poland in the years 1950–2021 (according to the 2019 zone classification) (Source: [24] based on data from general censuses from 1950 to 2021).

centers stagnate throughout the period, with no significant population change in 1960–2021 [24].

After the Second World War in Poland, internal population migrations were dominated by migrations from villages to cities. A similar tendency also occurred in the early 1980s. However, at the end of the 1990s, the direction from the city to the countryside began to predominate in internal migrations in Poland [7, 11], the essence of which was the appearance of movements from city centers in a centrifugal direction on an unprecedented scale—to extensively developed external zones (the so-called „internal suburbanization") or beyond the city's administrative borders [7, 11]. As a result, over the last three decades, the growth of suburban zones in Poland has been one of the most characteristic demographic and spatial processes [7, 11, 25]. With a fairly balanced level of registered internal migrations—approximately 400,000 events per year—there were significant changes in the structure according to area categories. The share of inflows was growing—from 17% (1989) to 34% (2021), with the fastest increase taking place until 2010, then the values remained at around 30%, to increase again by several points after 2015 percentages (Fig. 2) [11].

Suburbanisation processes substantially decrease the effectiveness of transport in European cities. This is related to growing dependence on car usage in those areas. Processes of suburbanisation change monocentric city areas into complex polycentric conurbations. Increasingly greater spatial-functional segregation is an important consequence of this process, resulting in making apartments, places of work, of trade, and of other services more distant [23, 26]. The suburbanisation and concentration of eco-

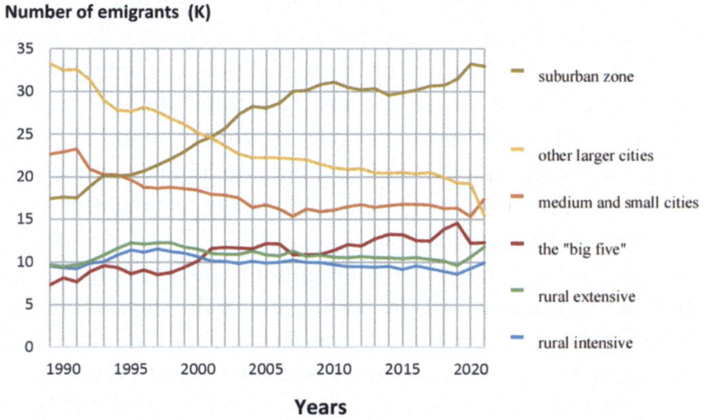

Explanations of the types of cities and municipalities:

- suburban zones: 456 communes around cities with over 50,000 inhabitants. Inhabitants;
- other larger cities: other voivodeship capitals and cities with over 50,000 inhabitants (total 82 cities);
- medium and small cities: other urban (49) and urban-rural (215) communes with a city of over 3,000 inhabitants. inhabitants (and less in the case of strongly developed non-agricultural functions);
- "wielka piątka" (red) - the "big five": Warszawa, Kraków, Poznań, Trójmiasto (Gdańsk, Gdynia Sopot), Wrocław;
- typically rural communes - rural communes (1,259) and urban-rural communes (195) outside suburban zones (in urban-rural communes, cities generally have fewer than 3,000 inhabitants); they were divided into "intensive rural" and "extensive rural" depending on the degree of development of the agricultural function and the share of poorly developed areas (forests, protected areas): rural extensive and rural intensive.

Fig. 2. Share in registered inflows in Poland in 1989–2021 of different types of areas (municipalities) (Source: [11] based on migration matrices of Statistics Poland).

nomic and service activities (e.g. shopping malls), which enable obtaining economies of scale, lead to a growing average length of travel and to travelling from one suburb to another. In general, suburbanisation consists in increased travel distances and thereby in being more dependent on a car [23]. Selected effects of suburbanization on the urban and suburban transport system and the mode of movement are presented in Fig. 3.

The ongoing processes of suburbanisation result in one more effect for the functioning of cities. The city development and selected functions frequently begin to go beyond its boundaries to adjacent administrative units. This may be a situation, in which the continuity of spatial investing is retained and the development is situated in so-called nearer neighbour zone, but it may also be carried out outside this zone, at a significant distance from the central city. City boundaries to a limited degree consider the proceeding suburbanisation processes; they were frequently determined many years ago, along the then existing lines of urban development. The fact, that in many cases the city boundaries do not fit to the city functioning, separate it artificially from the areas functionally linked with the city, causes also a problem for the urban public transport. However, the changing of boundaries is not only a formal issue, but it has also economic and social effects.

When organising the connections by the urban public transport it is difficult to close them within the administrative boundaries, because they reflect the demand resulting from the development and location of city functions, which due to the suburbanisation, is partly outside its administrative boundaries. So the integrated public transport must reflect the functional and spatial structure of the area as well as the places of residence; from the organisation point of view the administrative boundaries should be secondary in nature. This means, however, the necessity to determine the area of common organisation of the urban public transport, because the boundaries of the urban transport system do not enclose within the central city, agglomeration, or conurbation boundaries (Fig. 4). In this context the word 'urban' means a characteristic demand for services, short travel distances, travels in the morning hours starting from the place of residence to the place of work, education, or other activity, directional unevenness as well as the existence of morning and afternoon rush hours, and not the fact that transport lines and connections end within the city boundaries.

In the case of establishing organisers, the intention to integrate the public transport within the organiser activity area is just the basic premise for such a solution [27], and the determination of the optimal scope of integration is a decision related to the area of a specific organiser activity. This issues determine the transport policy goals and need to be reflected in the Sustainable Urban Mobility Plans (SUMPS) SUMPS are comprehensive, integrated mobility plans including all modes of transport. Many of them describe fare integration under conditions of suburbanization and include direct recommendations for spatial planning (for example SUMP for the city of Poznań or Beskid Agglomeration in Poland [28, 29]. For monocentric agglomerations the organiser activity may comprise only the big central city, but it can cover also satellite cities. In the case of polycentric agglomerations and conurbations the area of organiser activity comprises a few, a dozen or so, and even a few dozen cities, but it is always necessary to determine the boundary of the organiser activity. At the determination of the area, where solutions from the field of transport integration should be implemented, it is natural to consider the functioning of the city or metropolitan systems and the daily

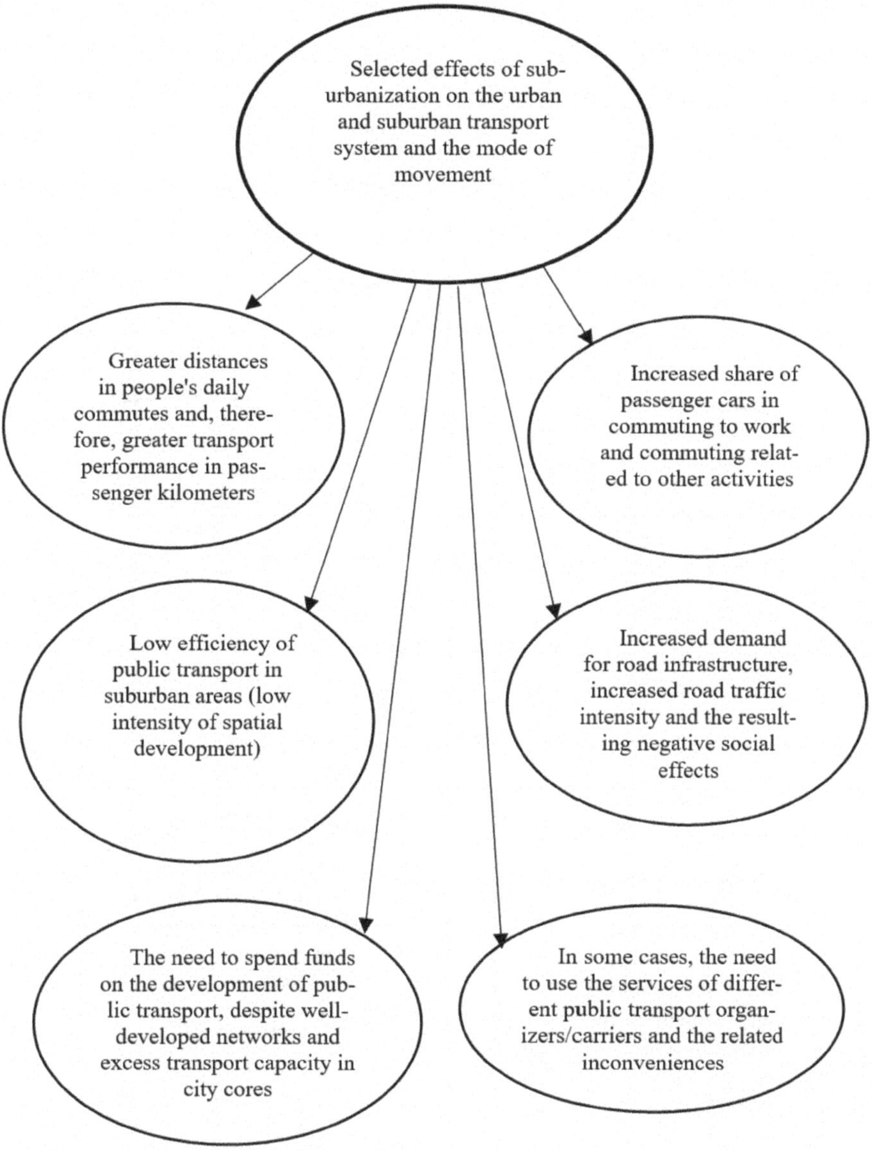

Fig. 3. Selected effects of suburbanization on the urban and suburban transport system and the mode of movement (Source: Own study).

movements of residents between municipalities that form them; it is necessary to determine boundaries of the area, which settlement units should be within one system of urban transport, especially that not always this area must be identical with the metropolis/agglomeration boundary. The analysis of municipalities diversification becomes important, because it will enable creating and adapting solutions taking into account the identity of some units. This may be done by means of taxonomic methods, that is meth-

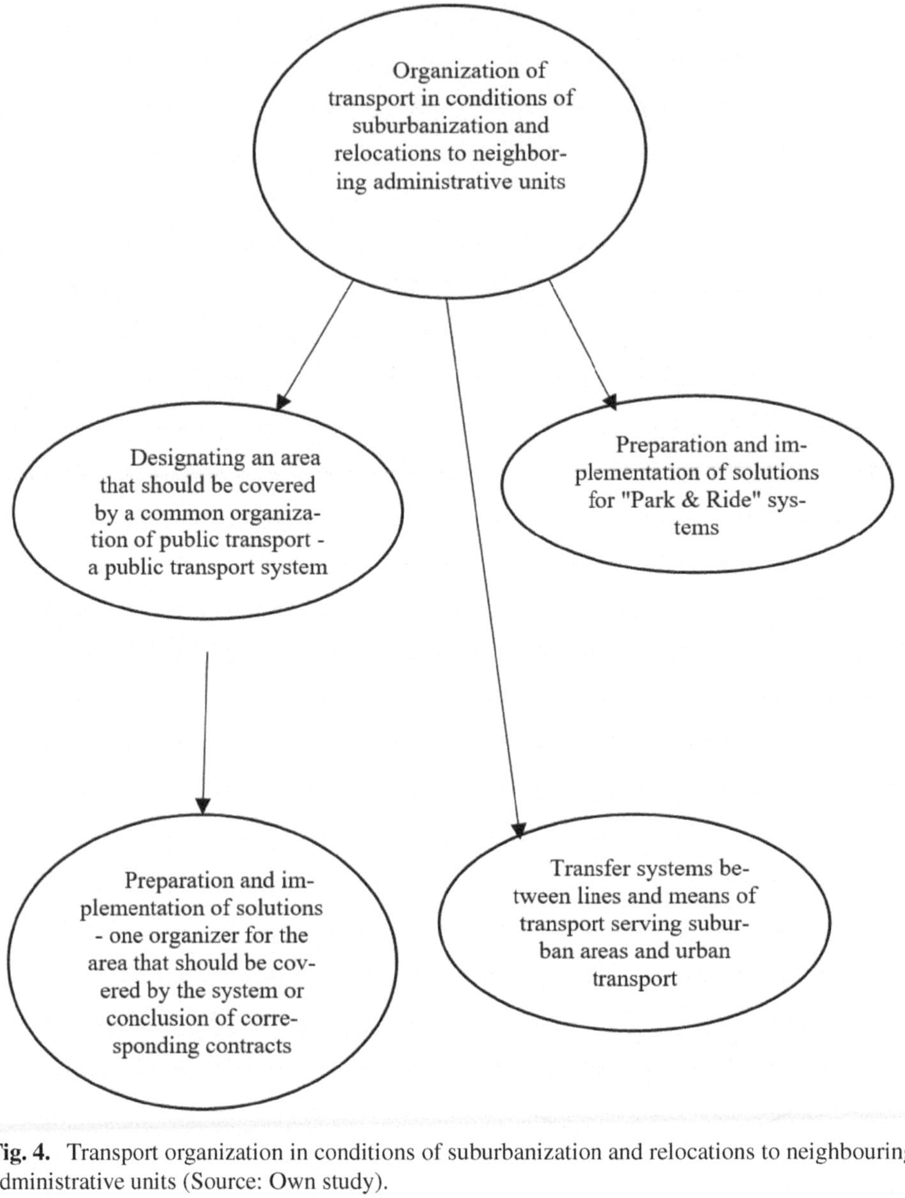

Fig. 4. Transport organization in conditions of suburbanization and relocations to neighbouring administrative units (Source: Own study).

ods used e.g. for classification of objects described by many properties, through determination of similarity between these objects. J. Czekanowski introduced quantitative methods to taxonomy as the first; he suggested a measure of similarity between objects in multi-dimensional space, he developed also a method for matrices ordering by a diagraphic method, referred to as the Czekanowski diagram. The Czekanowski method, originally proposed for anthropological studies, was and is used in many areas of biological, geographical, social, and economic studies [30]. This is a universal method of statistical classification, used in various fields of science, used also to merge territorial units, e.g. municipalities, to establish larger uniform regions [31]. The next methods in the group of taxonomic methods originated over time, including e.g. the Wrocław method and others.

The Czekanowski method was used in practice, making calculations for municipalities in the central part of the Silesian Voivodeship, to determine the area, where the public transport should be managed by one entity—the transport organiser [32]. The indication of the area was the starting point for the discussion on the planned solution; the final area covered by the uniform system is also an effect of political decisions, which also took into account the views and opinions of various social groups. In this solution the organiser sells tickets valid in the entire area, holds contracts with operators on the provision by them of transport services acc. to so-called gross cost formula. If within the organiser activity area also other entities are providing services, e.g. regional transport of the railway transport, it is in the organiser's interest to conclude contracts comprising at least the tariff integration within the area of its business, so as to expand in this way or just to achieve the situation, that all the offered transport services are covered by one system of sales, which will make services more attractive and will facilitate their use.

4 Differentiation of Fare Systems in Cities and in Suburban, Regional, and Inter-city Transport

In the public transport in large conurbations, both in Poland and abroad, a uniform or zone fare was dominating, which resulted from the expectations of a transparent and simple price list, and thereby ease of using the services of public transport. Simplification of the fare was pursued under conditions of mass urban transport. This resulted from the fact of carrying out sales mainly by external sales networks; maximum accessibility to tickets and a large number of points of sale were pursued in this way. However, this limited a possibility of introducing fares with a greater degree of complication. Also tickets are not always purchased for a specific ride, hence a uniform or zone fare gives greater possibilities during the purchase itself and, later on, during the selection of a connection, which can be used. The fare simplification affects also the minimisation of the cost of sales [33, 34]. The situation gradually changed with the introduction of electronic systems for fare collection, based on e-cards and, later on, successively with the use of mobile devices and applications dedicated to such equipment. The automated location systems used at that as well as databases on rides enabled to differentiate prices to a large extent, including also price charging with the use of a section-distance fare.

Enterprises carrying out transport of supralocal reach, e.g. transport from suburban areas, or of regional or national reach (bus or railway transport), most frequently apply a section-based fare. Obviously, there are also exceptions related to the transport in cities and the price adaptation to other carriers providing the transport services in a given area. Prices are diversified in terms of the rate for travelling on a specific section, and also of distance ranges in various companies. Degressive fares are applied, ticket prices converted to vehicle-kilometres go down with the increasing travel distance. Also diversification between individual enterprises very often occurs, part of enterprises diversify prices on the transport lines, too. This proves high flexibility and adaptation of prices to the specific market conditions.

For many years the diversification of fare systems in the urban and regional transport was one of barriers to the fare integration. It complicated the designing of ticket types used within an integrated offer and, in addition, it could be the reason for increased complexity of settlement algorithms in integrated distribution solutions. In recent years, with the popularisation of mobile devices and using them for fares payment, the importance of the barrier of various fare systems has decreased; it is possible to use, without excessive complications for passengers, various parallel fare systems, selected acc. to criteria set previously or by the passenger him/herself during a specific ride. However, another barrier has appeared, IT systems for public transport management, frequently including sales systems, implemented at various periods and by various suppliers; therefore their modifications are pretty labour-consuming and hence costly, so at to ensure relevant data exchange and integration of sale modules of various organisers and carriers, e.g. a system of an urban public transport organiser and of railway carriers. Also various approaches are used in this field for sales via mobile devices, organisers and carriers develop proprietary applications or use external suppliers; the function of paying the fares for the public transport services can be available in banking applications, too.

Another issue, with a social dimension, encountered during the integration of public transport systems, consists in diversified entitlements to concessionary and free travels; various legal regulations are in force in this case as well as solutions, which have developed during many years.

Solutions in the field of transport integration should be initiated by enterprises of regional, national, or international transport; such systems should frequently integrate not with one, but with several sub-systems of urban transport. Moreover, such carriers should be particularly interested in integration with the urban transport, because in a natural way the accessibility of railway stops as well as stations and airports is much smaller, than that of bus and tram stops of the urban transport. The offering of one ticket means the increasing of availability of railway or air transport, and hence easier using of those branches of transport. If we consider that the railway transport features a substantial share of fixed cost in the total cost, then the increase in the volume of transport means an increase in the income, and at the same time it does not generate significant additional costs. Also the fact of a high capacity of means of transport and very often non-utilisation of the rail or air infrastructure capacity should be added. Moreover, the inclusion of railways in the common ticket systems, and thereby an increased volume of railway transport, results in the improved efficiency of the entire public transport system and in decreased adverse environmental effects of the transport. In the case of higher occupancies of the rail transport a reduction of transport cost per passenger is achieved as well as external benefits related to the use of electric drive, and also reduction of roads load, which is very important in cities.

For an entity the fare integration may mean benefits, however, on the other hand the integration means limitation of entities independence, especially in so important decision area as the level of prices and sales, as well as numerous risks and concerns. The uncertainty and concerns are intensified by the fact that integration—cooperation is to take place with an entity, which was a competitor and, depending on the formulated rules of cooperation, still remains a competitor, e.g. on routes or in areas, which are not covered by the shared solutions. There are situations, in which entities unwillingly share the information on the technical and organisational knowledge of the carried out business, unwillingly cooperate, and this hinders the formation of integrated mobility services [35]. The projects and the scope of integration itself develop as a state of equilibrium between entities, affected by external regulations, expected benefits, estimated costs, concerns, and risks. Actions in the field of integration are also related to strategies for development and promotion, innovation, and social responsibility. So it is necessary to take into account not only solutions in the field of technologies oriented towards integration and services supply, but to assess mutual relationships between various interested parties, their motivations, and limitations [36]. Integrated fares combine a large consumption flexibility for the consumer, who can choose among many products at zero marginal monetary price, with the possibility of exploitation of the consumer's willingness to pay for the supplier [37].

The integration may comprise all lines and the entire area of activity of participating entities, and in the case of concerns, that the risk related to the correctness of division of funds originating from ticket sales is too high, an intermediate solution may consist in limitation of the scope of common tickets validity. The scope of common tickets may be limited only to selected types, e.g. 24-h or daily, or on the lines and routes to the railway station, airport, transfer centre, or another place significant from the point of view of transfers and of using various means of transport. As a result of scope reduction of fare integration, the amount of income on common tickets obtained in a specific period and divided between parties goes down, which also reduces the risks and financial effects related to the adopted method of income division. It is necessary to add, that the availability of common tickets in the case of part of people will mean giving up purchase of separate tickets, thereby for those persons smaller expenditures (a lower price of the common ticket than the sum of separate tickets), and for entities providing transport services—lower incomes. Of course, easier information searching and a possibility to buy a common ticket for different systems cause that the offered services are more attractive [36]; as a result, it is possible to expect an increase in the number of people using the public transport services [38, 39]. This increase, as it results from assessments carried out in various countries and cities, occurs after the introduction of integration solutions, however, it is noticeable only after at least a few months and usually lasts a few years. In general, the increase in the demand is estimated at a few to 12 or even 15% [40], but longer-term, i.e. after more than 20 years, figures of 20+ or even 30+ percent are reported [41].

5 Tickets for Lines Leading to High-Speed Railway Stations or Airports

In the case of starting lines to railway stations, ensuring intercity or international or airports connections, the basic risk consists in obtaining a relevant number of passengers using those lines, so as to achieve the intended level of payments. The ensuring of connections by means of urban public transport to the airport is frequently an ambition of local and regional authorities, and the provision of services is considered in the context of city image and promotion, as well as of the public transport systems itself. However, the number of passengers travelling by public transport and the income obtained on the ticket sales may turn out to be lower than originally planned, and the amount of subsidies significant. The airport authorities and air carriers are not interested in co-financing the lines ensuring travelling by the public transport to the airport. It means then the necessity to subsidise those lines from the public funds, i.e. by territorial self-governments. When travelling, in particular to the airports, which are situated in suburban areas, the use of cars, be it private or taxies, is common. The airport managers frequently offer car parks or groups of car parks, facilitating short-term parking related to driving persons to the airport, or leaving a car in long-term car parks, for the entire travel period, until a given person comes back. Apart from airports, car parks are offered also by other entities, situated in the vicinity of the airport, for whom this is a form of revenue earning. It is also necessary to add, that travelling to the airport by car is more convenient, especially when travelling with luggage, the travel time is shorter than by the public transport (excluding situations, where airports are connected by means of metro or railway systems); the travel start may be matched to the planned time of arrival at the airport. Unfortunately, in the case of large urban areas getting to the airport by means of public transport may require a few transfers; lines reaching the airports, to shorten the travel time, usually have only a few stops—in a few central points of the city or a group of cities.

The integration of sales or offering integrated tickets, for travelling to the airport and stations by the urban public transport and by air and rail transport, is a solution leading to the increase in the number of persons using the public transport lines. The following solutions are possible [42]:

- the organiser of urban public transport sells tickets for lines leading to stations and airports in the sales systems of railway transport or in sales systems of air tickets,
- the sale of tickets for urban public transport is available as an additional service/ product, which is provided by railway or air carriers,
- tickets entitling to travel by railway or air lines enable also travelling by selected lines of the urban public transport and/or for the adopted time before departure and after the arrival of a plane—for example the air carrier buys a specified number of seats in means of land public transport, which are next handed over—sold to passengers.

Unfortunately, in most cases, in which the aforementioned solutions have been implemented, the volume of sales of integrated tickets, enabling travelling by air and by public transport, is not high [42]. This results from several reasons. In the case of low-coast airlines, selling also other services and additional products, they are not interested in selling tickets for the public transport. Prices of public transport tickets, as against other services or products sold in air planes, are low, making that they are not a sales priority. At the same time commissions on sales, expected by the airlines, may reach even 40% of the ticket price. For the organisers and carriers of the urban public transport such commissions are high, substantially exceeding commission given to stationary points of sale, hence difficult to be accepted [41]. As a result, the organisers of the urban public transport carry out sales in their own sales networks, stationary points, ticket machines, or via the Internet, however under their systems of sales.

Despite limitations mentioned earlier, the solution ensuring integrated tickets of urban public transport and air transport has its growth potential [42]. It ensures travelling without the need to purchase separate tickets in the 'door-to-door' system, this is also an element of competitiveness improvement, especially for the air transport, against fast railway connections. Because airports are outside cities and railway stations most often in city centres. Another problem, when using the public transport travelling to the airport, is the fear that the travel to the airport will be delayed or cancelled, which would result in problems related to being late for the plane departure. The solution consisting in so-called travel guarantee is to be considered, hence in the case of non-performance of the travel or its significant delay, it is the public transport organiser, who together with the airline takes over the responsibility for the delay during the travel and ensures a possibility to fly, e.g. by the next plane. Also a solution, in which the air carrier takes over responsibility for delays during travelling by public transport to the airport, is possible. The interest in the public transport as a method to reach airports would increase, if the rules are similar as in the case of air tickets, in which from the start point to the endpoint of the travel a few flights are made [42]. In the case of delay in the first and next flight, the airline is obliged to ensure the travel continuation and, if necessary, to change the air ticket validity and the flight plan itself [43]. Unfortunately, these regulations do not apply to travelling by public and later on air transport, which results in passengers fears, and the carrier responsibility, if exists, is at its discretion.

Another challenge consists in developing, between air carriers and organisers of the urban public transport, standards of the exchange of data on fares, but also about the offer, and departure and arrival times. Because the implementation of solutions from the field of fare integration requires the integration of transport offer, hence such drawing of public transport timetables, to link them with planes departures and arrivals. This is a necessary condition for actions from the field of fare integration; without the timetables integration there will be no relevant interest in using lines of urban public transport, and the fare integration will not make much change. Charter flights may be an example, which substantial part in the summer period was taking place at night-time, hence at the time, when the offer of public transport to the airport is limited. In such a situation there are no premises to introduce combined tickets for the public transport and charter flights.

6 Premises and Factors of Fare Agreements Conclusions

Fare agreements are contracts concluded to unify the whole or part of fare regulations by entities having decision competence in the field of public transport. Fare agreements are generally concluded, when longer term preservation of separate identity of parties to the agreement is planned, pursuing at the same time unification of such an important element of service sales, as the fare. Agreements—contracts regarding the implementation and operation of common tariff systems are flexible solutions and can be implemented in various conditions (Fig. 5). For example, the solutions implemented based on the fare agreement can apply to:

- expansion of the area, in which the specific fare solution and a uniform ticket system are in force, by the next areas—municipalities or intermunicipal unions,
- unification of solutions and implementation of a uniform ticket in the entire offer and on all lines of urban and regional transport (various transport branches) or implementation of a solution limited to selected lines or relations,
- giving tickets for travelling in domestic and international relations by means of railway or air transport entitlements enabling the use of travels to stations by lines of local public transport.

The integration may comprise all lines and the entire area of activity of participating entities, and in the case of concerns, that the risk related to the correctness of division of funds originating from ticket sales is too high, an intermediate solution may consists in limitation of the scope of common tickets validity. The scope of common tickets may be limited only to selected types, e.g. 24-h or daily, or on the lines and routes to the railway station, airport, transfer centre, or another place significant from the point of view of public transport integration. As a result of scope reduction, the amount of income on common tickets obtained in a specific period goes down, which also reduces the risks and financial effects related to the adopted method of income division. It is necessary to add, that the availability of common tickets in the case of part of people will mean giving up separate tickets, thereby for those persons smaller expenditures (a lower price of the common ticket than the sum of separate tickets), and for entities providing transport services—lower income. It is obviously expected that longer-term the number of persons using the public transport will increase, however longer-term the number of persons using the services is also affected by other factors. As a result, the benefits of integration in the form of an increased number of persons using the public transport and additional revenue from sales of services are pretty difficult to estimate in practice.

Another solution from the field of integration consists in extension of the area, in which a specific fare solution and a uniform ticket system are in force, by the next areas—municipalities or intermunicipal unions. This applies both to the common lines, i.e. connecting the organiser's area with neighbouring municipalities, or all lines of parties to the agreement. In this case there are no limitations of various statutory regulations in the field of entitlements to concessionary or free rides. The agreement concluded between the parties requires that the rules of income on tickets division should be determined. In a similar way as in the case of solutions in the field of common tickets for the urban public transport and carriers providing transport on suburban, regional, or national lines, the development of a relevant method to determine the revenue obtained on tickets sales is a challenge and also a risk. The source of various opinions, and hence the risk of correctness of the adopted methods of division and settlements, consists in the determination to what extent, in the case of lines passing through the area of various municipalities/organisers, the shortage on the entire transport line is averaged, and to what extent the fact is considered, that their uneven occupancy can exist, and frequently

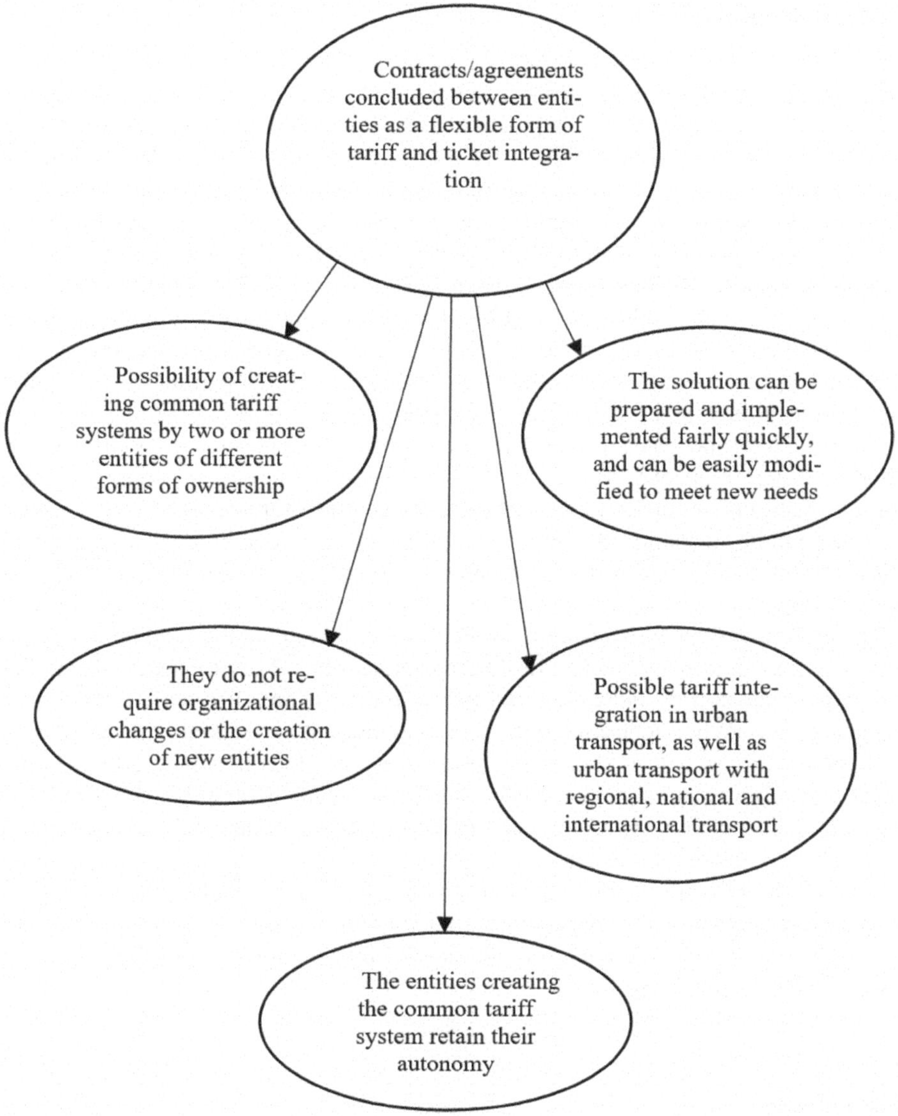

Fig. 5. Contracts/agreements concluded between entities as a flexible form of fare and ticket integration (Source: Own study).

exists, on the route. Lines connecting suburban areas and smaller municipalities with the centre of a big city are characterised by the fact, that the number of passengers increases with approaching the city centre, and that means a decline of deficit on a given line, if it serves the area, in which it carries a higher number of passengers. In addition, the transport in city centres features a smaller distance of travel, which also affects the amount of generated shortages.

7 Conclusions

The processes of suburbanisation proceed not only within administrative boundaries of a given city, but result in the development of areas in administration units neighbouring the city. They have significant effects for transport systems, result in larger distances between the travel origins and destinations, and hence a greater transport intensity. In addition, a low density of development and its scattering results in the lack of concentration on main routes and thereby their extension, as well as lower utilisation of means of public transport. The development originates, to which people commute mainly by cars, because in such conditions it is difficult to achieve the quality of public transport services, which is competitive to travelling by cars. At the same time the population of city centres goes down, hence in the areas with a well-developed network of public transport, including in particular the rail transport.

Irrespective of the very effects related to the location and covering the distance, the suburbanisation frequently also results in using, when commuting to central city areas, various organisers or carriers; there are different centres of transport services management and public financing. Because commuting to cities combine the use of public transport with a suburban or regional nature (e.g. railway connections) and of urban public transport.

The processes of suburbanisation, as well as location of stations and an increase in the importance of airports situated in suburban areas, should release actions in the field of transport integration, individual and public transport and also various systems of public transport. When focusing on the fare integration it is necessary to state, that the reach of organisers action should not necessarily be determined by administrative boundaries of units, but should consider daily movements of suburban areas residents as well as the location of the aforementioned infrastructure of national and international reach. The conclusion of fare agreements is another solution. The solutions of common tickets, which are being implemented now, can use IT, in particular payment systems using mobile applications; they become a tool not only for service promotion and sales, but also, together with solutions from the field of information and planning, a method to shape the transport behaviour.

References

1. Le Néchet, F.: Urban spatial structure, daily mobility and energy consumption: a study of 34 European cities. Cybergeo: European Journal of Geography [Online], Systems, Modelling, Geostatistics, document 580, Online since 23 January 2012, connection on 27 August 2022. https://doi.org/10.4000/cybergeo.24966 (2012)
2. Pelosi, A.: The Comparative Evolution of Suburbanization. Major Research paper, University of Guelph. https://atrium.lib.uoguelph.ca/xmlui/bitstream/handle/10214/16201/PelosiA_201512_MRP.pdf?sequence=1&isAllowed=y (2015)
3. Hesse, M., Siedentop, S.: Suburbanisation and suburbanisms – making sense of continental European developments. Raumforsch Raumordn Spat Res Plan. **76**, 97–108 (2018). https://doi.org/10.1007/s13147-018-0526-3

4. Urban Sprawl in Europe: The ignored challenge. European Commission Directorate – General Joint Research Centre, European Environment Agency, No 10/2006, ISSN 1725-9177 (2006)
5. Różycka-Czas, R., Czesak, B., Staszel, A.: Which Polish cities sprawl the most. Land. **10**(12), 1291 (2021). https://doi.org/10.3390/land10121291
6. Śleszyński, P., et al.: Suburbanizacja w Polsce jako wyzwanie dla polityki rozwoju. (2022). https://doi.org/10.24425/140305
7. Kurek, S.: Territorial distribution of population change in Poland in the years 1991-2001. Bull. Geogr. Socio-Econ. Ser. **4**, 117–134 (2005)
8. Kurek, S., Rachwał, T., Wójtowicz, M.: Industrial and commercial suburbanization in post-socialist city: the Kraków Metropolitan Area (Poland). Annales Universitatis Paedagogicae Cracoviensis Studia Geographica. **5**, 55–76 (2014)
9. Kurek, S., Wójtowicz, M., Gałka, J.: Functional urban areas in Poland: demographic trends and migration patterns. (2020). https://doi.org/10.1007/978-3-030-31527-6
10. Śleszyński, P.: Demograficzny wymiar procesów suburbanizacji w Polsce po 1989 roku. In: Kozłowski, W.S. (ed.) Żywiołowe rozprzestrzenianie się miast. Narastający problem aglomeracji miejskich w Polsce. Studia nad Zrównoważonym Rozwojem, pp. 105–124. Ekonomia i Środowisko, Białystok-Lublin-Warszawa (2006)
11. Dolińska, A., et al.: Migracje z miast na wieś: determinanty oraz wybrane konsekwencje w wymiarze ekonomicznym i przestrzenno-środowiskowym : (na przykładzie strefy podmiejskiej Wrocławia) Migracje z miast na wieś. Determinanty oraz wybrane konsekwencje w wymiarze ekonomicznym i przestrzenno-środowiskowym (na przykładzie strefy podmiejskiej Wrocławia), 2023, Wydawnictwo Uniwersytetu Ekonomicznego we Wrocławiu, ISBN 9788367400411, [9788367400428] (2023)
12. Kopecky, K.A., Suen, R.M.H.: A quantitative analysis of suburbanization and the diffusion of the automobile. Int. Econ. Rev. **51**(4) (2010) http://karenkopecky.net/SandAPaper.pdf
13. Yukawa, S.: Suburbanization and urban public transport. Declining public transport in Japanese regional city and regional transport policy, 11th Conference on Competition and Ownership in Land Passenger Transport Delft University of Technology (The Netherlands), 20–25 September 2009. https://www.researchgate.net/publication/277859349_Suburbanization_and_Urban_Public_Transport_Declining_public_transport_in_Japanese_regional_city_and_regional_transport_policy (2009)
14. Kruszyna, M., Śleszyński, P., Rychlewski, J.: Dependencies between demographic urbanization and the agglomeration road traffic volumes: evidence from Poland. Land. **10**(47) (2021). https://doi.org/10.3390/land10010047
15. Narefsky, K.: The Suburbanization of the US Working Class. https://jacobin.com/2014/10/the-suburbanization-of-the-us-working-class (2014)
16. Allen, J., Farber, S.: Suburbanization of transport poverty. Ann. Am. Assoc. Geogr., 1–18 (2021). https://doi.org/10.1080/24694452.2020.1859981
17. Zimmerman, R., et al.: Suburban poverty, public transit, economic opportunities, and social mobility. University Transportation Research Center - Region 2, December 2015, UTRC/RF Grant No: 49997-25-25. http://www.utrc2.org/sites/default/files/Final-Report-Surburban-Poverty-Public-Trans-Eco-Opportunities.pdf (2015)
18. Anas, A., Arnott, R., Small, K.A.: Urban Spatial Structure, May 1, 1998. https://www.socsci.uci.edu/~ksmall/JEL%20Paper.pdf (1998)
19. Taubenböck, H., et al.: Seven city types representing morphologic configurations of cities across the globe. Cities. **105**, 102814 (2020). https://doi.org/10.1016/j.cities.2020.102814
20. Stangel, M.: Kształtowanie współczesnych obszarów miejskich w kontekście zrównoważonego rozwoju, p. 69. Wydawnictwo Politechniki Śląskiej, Gliwice (2013) https://delibra.bg.polsl.pl/Content/20828/REPO_22455_-_Ksztaltowanie-wspolc_0001.pdf

21. Bontje, M., Burdack, J.: Edge cities, European-style: examples from Paris and the Randstad. Cities. **22**(4), 317–330 (2005). https://doi.org/10.1016/j.cities.2005.01.007
22. Heffner, K.: Proces suburbanizacji a polityka miejska w Polsce. In: Marszał, T. (ed.) Miasto – region – gospodarka w badaniach geograficznych. W stulecie urodzin Profesora Ludwika Straszewicza, pp. 75–110. Wydawnictwo Uniwersytetu Łódzkiego, Łódź (2016). https://doi.org/10.18778/8088-005-4.05
23. Dydkowski, G.: The Impact of Cities' Spatial Planning on the Development of a Sustainable Urban Transport, pp. 13–25 (2019). https://doi.org/10.1007/978-3-030-35543-2_2
24. Śleszyński, P.: Prawidłowości i nieprawidłowości rozwoju stref podmiejskich w Polsce, w: Suburbanizacja. Człowiek – zmiana – przestrzeń, Materiały z III Kongresu Demograficznego. Część 15, Rządowa Rada Ludnościowa, Warszawa 2023. https://kd.stat.gov.pl/images/publikacje/02_Suburbanizacja.%20Warszawa%20III%20kongres.pdf (2023)
25. Jończy, R., et al.: Environmental and economic factors of migration from urban to rural areas: evidence from Poland. Energies. **14**(24), 8467 (2021). https://doi.org/10.3390/en14248467
26. Vanoutrive, T.: Commuting, spatial mismatch, and transport demand management: the case of gateways. Case Stud. Transp. Policy. **7**(2), 489–496 (2019). https://doi.org/10.1016/j.cstp.2018.12.011
27. Tomanek, R.: Strategies of transport horizontal integration and their influence on the competition. The Karol Adamiecki University of Economics in Katowice. J. Econ. Manag. **4** (2008) https://www.ue.katowice.pl/fileadmin/_migrated/content_uploads/9_R.Tomanek_Strategies_of_transport_horizontal....pdf
28. Plan Zrównoważonej Mobilności Miejskiej dla Aglomeracji Beskidzkiej 2040+. https://aglomeracjabeskidzka.eu/poziom-lokalnysubregionalny
29. Plan Zrównoważonej Mobilności Miejskiej dla Metropolii Poznań do 2040 roku. https://aglomeracjabeskidzka.eu/poziom-lokalnysubregionalny
30. Pociecha, J.: Rozwój metod taksonomicznych i ich zastosowań w badaniach społeczno-ekonomicznych, Materiały z konferencji "Statystyka społeczna. Dokonania – szanse – perspektywy". GUS, Warszawa (2010) https://stat.gov.pl/cps/rde/xbcr/gus/POZ_Rozwoj_metod_taksonomicznych.pdf
31. Szymla, Z.: Determinanty rozwoju regionalnego. Ossolineum, Wrocław (2000)
32. Dydkowski, G., Kos, B.: Metropolizacja zarządzania miejskim transportem zbiorowym w województwie śląskim, Studia Ekonomiczne nr 336/2017 Uniwersytet Ekonomiczny w Katowicach, Katowice. https://www.ue.katowice.pl/fileadmin/user_upload/wydawnictwo/SE_Artyku%C5%82y_321_340/SE_336/01.pdf (2017)
33. Tomanek, R., et al.: Ceny transportu miejskiego. Akademia Ekonomiczna w Katowicach, Katowice (2007) ISBN 9788372468291
34. Dydkowski, G., Tomanek, R., Urbanek, A.: Taryfy i systemy poboru opłat w miejskim transporcie zbiorowym. Wydawnictwo Uniwersytetu Ekonomicznego w Katowicach, Katowice (2018) ISBN 978-83-7875-440-4
35. European Commission: Sustainable and Smart Mobility Strategy – putting European transport on track for the future. Communication from the Commission to the European Parliament, the Council, the European Economic and Social Committee and the Committee of the regions, COM(2020) 789 final, Brussels, 9.12.2020. https://eur-lex.europa.eu/legal-content/EN/TXT/?uri=CELEX%3A52020DC0789 (2020)
36. Puhe, M.: Integrated urban e-ticketing schemes – conflicting objectives of corresponding stakeholders. Transp. Res. Procedia. **4**, 494–504 (2014). https://doi.org/10.1016/j.trpro.2014.11.038
37. Marchese, C.: The economic rationale for integrated tariffs in local public transport. Ann. Reg. Sci. **40**(4), 875–885 (2006). https://doi.org/10.1007/s00168-005-0045-3

38. Abrate, G., Piacenza, M., Vannoni, D.: The impact of Integrated Tariff Systems on public transport demand: evidence from Italy. Reg. Sci. Urban Econ. **39**(2), 120–127 (2009). https://doi.org/10.1016/j.regsciurbeco.2008.05.014
39. Booz & Company: The Benefits of Simplified and Integrated Ticketing in Public Transport. Passenger Transport Executive Group, United Kingdom, October 2009. https://www.urbantransportgroup.org/system/files/general-docs/integratedticketingreportFINALOct09.pdf (2009)
40. Puhe, M. et al.: Integrated urban e-ticketing for public transport and touristic sites. Science and Technology Options Assessment. Report PE 513.551. ISBN 978-92-823-5302-8, https://doi.org/10.2861/49821 (2014)
41. NEA: Integration and Regulatory Structures in Public Transport. Final Report. Project leader: NEA Transport research and training. European Commission DG TREN, Rijswijk (2003)
42. Surface Access: Integrated Ticketing Report, North Star Atkins. https://assets.publishing.service.gov.uk/government/uploads/system/uploads/attachment_data/file/763039/surface-access-integrated-ticketing.pdf (2018)
43. Regulation (EC) No 261/2004: Regulation (EC) No 261/2004 of the European Parliament and of the Council of 11 February 2004 establishing common rules on compensation and assistance to passengers in the event of denied boarding and of cancellation or long delay of flights, and repealing Regulation (EEC) No 295/91 (Text with EEA relevance) - Commission Statement. https://eur-lex.europa.eu/legal-content/PL/TXT/?uri=celex%3A32004R0261 (2004)

Empirical Investigation into the Evaluation Methodology for Transport Interchange Accessibility for Individuals with Special Needs: An In-Depth Analysis of Selected Transport Interchanges in Gdańsk

Romanika Okraszewska[1] (✉), Łukasz Jeliński[1], Bartosz Nowak[1], and Kazimierz Jamroz[1]

[1] Gdansk University of Technology, Gdańsk, Poland
romanika.okraszewska@pg.edu.pl; lukasz.jelinski@pg.edu.pls;
s186265@student.pg.edu.pl; kazimierz.jamroz@pg.edu.pls

Abstract. The manuscript delineates an empirical research methodology for gathering data for the functional characterization of transport interchanges. The adoption of a universal, standardized field research method ensures the feasibility of objective interchange comparisons. The explicated methodology encompasses observational and survey research, yielding both quantitative and qualitative data. The amassed data comprises the count of individuals utilizing integrated interchanges, the tally of individuals embarking and disembarking from each mode of transportation, the count of individuals transitioning at the interchange, and the tally of passengers with special needs. Qualitative data encompasses evaluations of interchanges by individuals with special needs, transportation behavior patterns, motivations, and the origins and destinations of travel. The collated data can also aid the interchange manager in decision-making processes pertinent to the maintenance and development of the interchange. The methodology incorporates initial assumptions, data collection methodologies, tools (such as survey sheets and tables), and data analysis techniques.

Keywords: Interchange transport nodes · Accessibility · Empirical investigation · Passengers with special needs

1 Introduction

1.1 Background

Interchanges (transport interchanges, integrated interchanges, transport integration hubs, transfer centers, or integrated stops) play a pivotal role in the functionality of the multimodal passenger transport system within urban environments. They serve as critical nodes for managing passengers boarding and alighting from various transport modes and facilitate the transition of travelers between modes or directions.

G. Sierpiński et al. (Eds.): TSTP 2024, LNNS 1257, pp. 179–197, 2024.
https://doi.org/10.1007/978-3-031-81777-9_12

A significant proportion of existing nodes necessitate enhancements in the integration between modes of passenger transport and adaptations to cater to the requirements of individuals with special needs (PwSN). Moreover, there is a need to augment the consistency of the solutions implemented. Consequently, within the ambit of the 'Interchange without Barriers' initiative, the challenge of augmenting the accessibility of interchanges was addressed by formulating a principle of consistency applicable to all components of an integrated interchange, with a particular focus on all categories of individuals with special needs.

An interchange comprises several constituent elements, specifically: routes leading to the interchange, passenger arrival points (stops, parking lots, boarding and alighting platforms), passenger terminals with departure points, and passenger movement pathways within the interchange vicinity. For the purpose of accommodating travelers with special needs, three elements of the interchange are of paramount importance: passenger terminals, passenger arrival points (public transport stops and parking lots), and passenger movement routes.

Adapting the definition of PwSN from the Polish Act [1], travelers with special needs are individuals who, due to their inherent or circumstantial characteristics, necessitate additional measures or resources to surmount a barrier in order to realize travel (interchange) on an equitable basis with other (able-bodied) individuals.

Identification of people with special needs and their classification is a complex and difficult task [2]. Based on the criterion of ability, which is crucial for fully, independently, and equitably undertaking activities related to travel, five characteristic groups of travelers have been identified: individuals in circumstances affecting mobility, individuals who are blind or visually impaired, individuals who are deaf or hard of hearing, individuals with limited skills in direct or indirect communication, and individuals with limited psycho-social abilities [3].

These diverse groups exhibit distinct needs and expectations of the solutions, services, and facilities offered. Failure to meet the needs of users and the resulting expectations can pose a hindrance or a critical barrier to the utilization of a transportation hub and, consequently, public transportation [4].

1.2 Aim and Scope of the Study

The methodology for conducting research on interchanges, as delineated in the manuscript, along with the presented findings and conclusions, are the culmination of work undertaken by a consortium of researchers from the Gdansk University of Technology. This was encapsulated within a research project titled 'System for Managing the Accessibility of Integrated Interchanges' ('Transfer without Barriers—TwB') [5]. A primary task of the project was to devise a method for assessing the accessibility of Integrated Transport Interchanges (ITI) for individuals with special needs (PwSN). As a sub-task, a preliminary research methodology was adopted to procure information characterizing the interchanges, and pilot studies were conducted on selected interchanges. The results of the research were utilized to characterize the nodes and develop the classification of transport interchanges, criteria for evaluation, and minimum accessibility requirements for each type of transport interchange [6, 7]. Conclusions drawn from the pilot studies were employed to develop the final methodology for conducting field research at transport interchanges.

The authors of the 'Method for Auditing Existing Integrated Interchanges' advocate the use of the developed method to obtain the required minimum (or optimal) information for the functional characteristics of the interchange. The information procured from the research can be additionally beneficial to the interchange manager and support the process of managing the accessibility of the ITI and the decision-making process for maintenance and development of the interchange. Recommended quantitative studies encompass measuring the number of people using ITI, the number of people boarding and alighting from each mode of transportation at ITI, the number of people transferring to ITI, the number of PwSNs using ITI, among others. Qualitative research includes PwSN's evaluation of ITI, identification of transportation behavior, motivations, sources, and purposes of travel. Information collected on passenger behavior, operation of ITI, or PwSN's evaluation of ITI equipment is also included.

To achieve a standardized assessment of accessibility, it is imperative to utilize reliable and comparable measurement data. Consequently, it is advantageous that research on nodes be conducted in accordance with a singular, universal methodology that ensures the repeatability and representativeness of the results. The manuscript introduces the methodology of conducting empirical research at the ITI, adopted as a result of the 'Transfer without Barriers' project implementation. The second chapter delineates the general assumptions of the methodology, the methods and types of research employed, and the methods of result analysis. The third chapter is dedicated to the method of analyzing research results. The fourth chapter provides an in-depth case study of two transfer interchanges in Gdańsk: Gdańsk Świętokrzyska and Gdańsk Śródmieście. The fifth chapter presents the most significant research results in a synthetic manner. The concluding chapter provides a summary and indicates directions for future research.

2 Research Method

The devised methodology presupposes the implementation of the following research methods:

Research predicated on video recording materials—This encompasses the observation and measurement of passenger and vehicle flow predicated on the recorded visual representation of the entire transfer hub.

Research predicated on on-site observation—This is conducted by pre-trained personnel and includes surveys with Individuals with Special Needs (ISN) and the enumeration of individuals boarding and alighting from the means of transport.

2.1 General Assumptions

The proposed methodology is designed to be universally applicable, facilitating research on any selected node in a clear and repeatable manner. This ensures the comparability of results across different nodes and measurement moments. The comprehensive research plan incorporates video recording techniques and field studies, which include conducting surveys and quantifying passenger numbers at the ITI.

In the preparatory phase, it is essential to delineate the interchange in terms of its geographical location, defined boundaries, its role in the local transport system, the type of node, and the integrated transport modes. Optionally, subject to data availability, additional information such as the volume of passenger traffic and parameters of elements and objects can be provided.

Field studies are to be conducted during a typical weekday (excluding holidays and non-working days), specifically from Tuesday to Thursday, during the academic year, and during the morning and afternoon peak hours. Video recording tests should cover a full 24-h period on a typical weekday. This rigorous methodology ensures a thorough and accurate understanding of the interchange's functioning and usage patterns.

2.2 Measurements of Passenger Flows

The quantification of passenger flows necessitates the enumeration of individuals entering and exiting the interchange node area, as well as those traversing the Integrated Transport Interchanges (ITI) within delineated measurement cross-sections, which are further subdivided into routes. The investigation can be partitioned into preparatory stages and main research, which encompasses four operations: the construction of an interchange diagram, the designation of screens, the installation of cameras, and the analysis of recordings.

The task may appear straightforward, however, the complexity escalates in proportion to the size and intricacy of the node's spatial structure. Consequently, the initial step involves defining the boundaries of the interchange, conducting an inventory to pinpoint the locations where passengers converge on the node (such as pedestrian entrances to the junction, stops, platforms, parking spaces, bicycle shelters, etc.), and identifying the elements and objects situated therein. This process also involves determining the movement routes that connect individual elements. The data gathered and plotted on the map form a **diagram of the interchange node**, serving as a foundation for subsequent work.

For the successful execution of the research, it is imperative to identify the origins and destinations of pedestrian movements around the interchange, as well as the locations where passenger flows diverge into individual directions. Within these locations, virtual areas, referred to as **screens** (analogous to a road intersection), are established. Within these screens, passenger flows are segregated according to the predetermined directions and relations. Individuals passing through these screens are counted at specified time intervals, and are categorized into routes and directions. The methodology provides a detailed explanation on how to ascertain relationships and directions, and includes examples of markings and descriptions (see Fig. 1).

Fig. 1. An illustration of a virtual screen delineates the main connections and directions of pedestrian mobility within the Gdańsk Świętokrzyska interchange node.

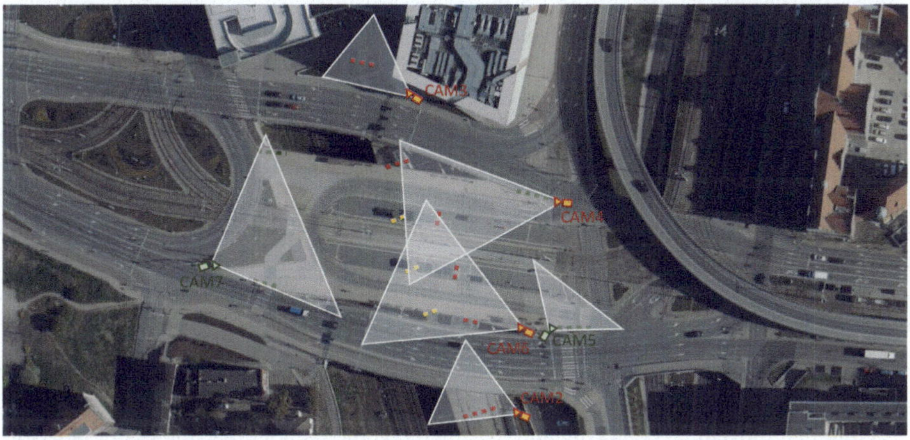

Fig. 2. An illustration of camera installation at the Gdańsk Śródmieście. These cameras are mounted on city lamps, with the explicit permission of the road manager.

A crucial phase of the preparatory work involves determining the quantity and positioning of **cameras**. The paramount criterion is to ensure comprehensive coverage of the entire interchange area, encompassing all entry and exit points to the study area, as well as significant intersections of pedestrian flows (see Fig. 2). The ultimate decisions are guided by factors such as the variety of cameras accessible, the feasibility of camera installation in relation to the physical infrastructure of the node, and the proprietary rights associated with the elements situated within its vicinity.

The primary phase of the research involves the **scrutiny of recorded footage** and the quantification of passenger flows. The responsibility of video material observers is to enumerate individuals traversing the screen area, segregated into distinct directions and relations, and to identify passengers with special needs. The cohort of interchange users excludes bus and tram operators transitioning between vehicles and facilities situated at the interchange (should such instances arise). To ensure the universality of the research, the methodology has devised a classification system for passengers with special needs groups to facilitate unambiguous identification of users. Additionally, a template of a universal measurement card has been developed, which is tailored to the requirements of the examined node in accordance with the established measurement screens and displacement relations.

2.3 Quantification of Passengers Boarding and Alighting Transport Mode

The process of monitoring the volume of passengers utilizing the transport services integrated within the interchange is conducted within the vicinity of the interchange by trained observers. These observers are tasked with enumerating the passengers embarking and disembarking individual transport services, in addition to assessing the occupancy level of vehicles entering the interchange node. This method encompasses a preparatory phase and a fundamental research phase. The preparatory phase involves training the observers and setting up the necessary infrastructure for data collection. The fundamental research phase involves the actual data collection and subsequent analysis. This methodology provides a robust framework for understanding passenger dynamics at interchange nodes, thereby informing strategies for efficient management and potential enhancements (Fig. 3).

Prior to initiating the measurement process, it is imperative to designate measurement stations predicated on the developed node diagram. Subsequently, an appropriate number of observers are assigned to these stations, taking into consideration the perceptual capabilities of an individual observer (see Fig. 4). Observers are mandated to conduct observations and measurements strictly within their assigned areas, adhering to the following protocol:

- record information pertaining to the approaching vehicle, which includes: arrival time, line number, vehicle side number, vehicle occupancy level (see Fig. 5)
- enumerate individuals entering and exiting the vehicles; identify and count individuals with special needs belonging to defined groups (see Fig. 5).

This systematic approach ensures accurate data collection and aids in the comprehensive analysis of passenger dynamics at interchange nodes.

2.4 Measurements of the Number of Vehicles and People Using Parking Lots

Vehicle flow research entails the enumeration of vehicles entering and exiting a parking lot, a process that is also based on recorded video footage. The study can be bifurcated into a preparatory stage and a primary research stage. The preparatory stage involves the development of a parking diagram and the installation of cameras. The primary research stage encompasses the investigation of vehicle flows and space occupancy, as well as the measurement of pedestrian traffic within the parking lot.

The initial stage constitutes an inventory of the parking lot, which includes determining the boundaries of the parking lot, the number of public parking spaces, the

Fig. 3. A method of visually assessing the filling of a means of transport.

number of spaces designated for disabled individuals, and the number and location of entrances/exits to/from the parking lot.

The complexity of conducting the research is contingent upon the size and organization of the parking lot. For instance, surface parking lots devoid of additional structures or large parking lots with a single entrance may necessitate only one camera. However, in the case of expansive or multistorey car parks with multiple entrances/exits, a sufficient number of cameras must be installed to ensure the recording of all entries and

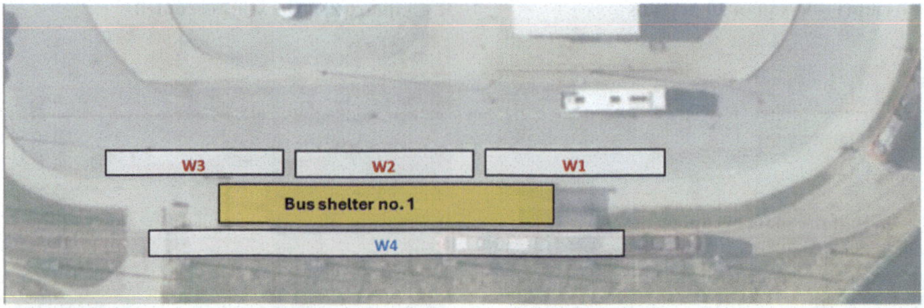

Fig. 4. An example of designating measuring stations at the transfer stop at the Gdańsk Świętokrzyska interchange node.

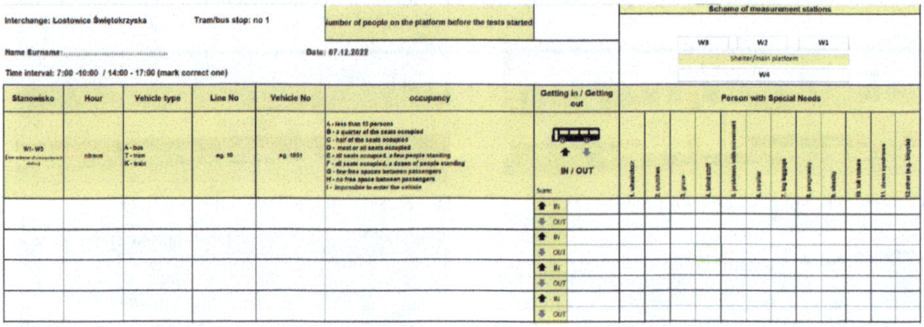

Interchange: Łostowice Świętokrzyska
Tram/bus stop: no 1
Number of people on the platform before the tests started
Name Surname:
Date: 07.12.2022
Time interval: 7:00 -10:00 / 14:00 - 17:00 (mark correct one)

Scheme of measurement stations: W3, W2, W1, Shelter/main platform, W4

Stanowisko	Hour	Vehicle type	Line No	Vehicle No	occupancy	Getting in / Getting out	Person with Special Needs											
							1. wheelchair	2. crutches	[illegible]	4. blind staff	5. problems with movement	6. stroller	7. big luggage	8. pregnancy	9. obesity	10. tall stature	11. down syndrome	12 other (e.g. bicycle)
W1 - W5		A - bus T - tram K - train	eg. 10	eg. 1001	A - less than 10 persons B - a quarter of the seats occupied C - half of the seats occupied D - most or all seats occupied E - all seats occupied, a few people standing F - all seats occupied, a dozen of people standing G - few free spaces between passengers H - no free space between passengers I - impossible to enter the vehicle	IN / OUT Sum:												
						IN OUT												
						IN OUT												
						IN OUT												
						IN OUT												

Fig. 5. An example of a form completed by an observer during measurements at a node.

exits. In certain scenarios, it may not be feasible to identify individuals traversing the parking lot (see Fig. 6).

The measurement results are recorded in a form tailored to the specific node under study. To optimize time efficiency, the material can be reviewed in an accelerated playback mode.

Counting individuals who traverse the parking lot, without it being their point of origin or destination, is an optional component of the study. This aspect is contingent upon the observers' resources, including time and financial limitations. Given the unique spatial configurations of each node, it is essential to develop a matrix of identified relationships. This matrix should include the count of individuals passing through, thus providing a comprehensive understanding of the node's dynamics.

2.5 Research on the Behavior and Transport Preferences of Passengers with Special Needs

The study of transport behavior and preferences is conducted using survey research methods. Interviews are carried out with the interchange's users to gather information on the transport behavior of individuals with special needs (PwSN) and their opinions on the technical condition of the node and its adaptation to their needs. Interviewers are

Fig. 6. An example of a parking lot inventory at the Gdańsk Świętokrzyska interchange.

tasked with identifying individuals with special needs and interviewing them using semi-structured questionnaires. These questionnaires include six closed questions regarding travel and one open question about potential barriers and issues encountered while traveling.

The closed questions address the following aspects: the starting and ending points of the trip, the purpose of the trip, the modes of transport used to reach and leave the junction, and an assessment of the junction's technical condition on a three-point scale. Additionally, during each interview, the interviewer records observations on the age group, gender, and any visible special features of the respondents. The survey sheet must be adapted to the specific conditions of each node, such as the districts that can be selected in questions about the trip's start and end locations.

The number of interviewers and their working hours are determined based on the expected outcomes of the study. Considering the number and nature of the questions, it is estimated that an interviewer can conduct a maximum of 12 interviews per hour.

3 Analysis of the Collected Data

3.1 Verification of Collected Data

In scenarios where there is no possibility of new individuals entering or exiting the passenger flow between cross-sections (e.g., a tunnel with a single entrance and exit), the values of people crossing adjacent cross-sections can be verified through comparative analysis. This concept is illustrated using the example of adjacent cameras X and Y in Fig. 7. The number of individuals passing through the NL, WW, and SP sections recorded by camera X should correspond to the number of individuals passing through the WL, WW, and WP sections recorded by camera Y.

3.2 Passengers Exchange Scheme

The exchange of passengers within a stop involves individuals entering the stop area from adjacent regions as well as from public transport vehicles arriving at the stop. It also includes individuals exiting the stop area, either on foot or by using public

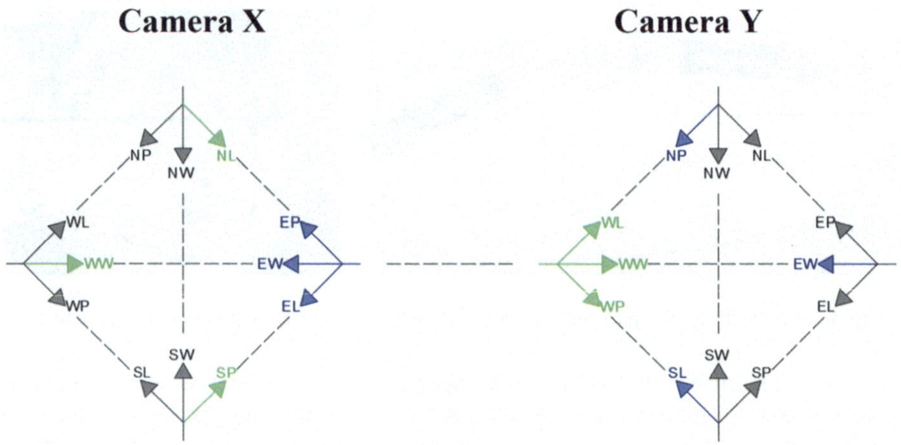

Fig. 7. A diagram explaining how to verify passenger number measurements in cross-sections.

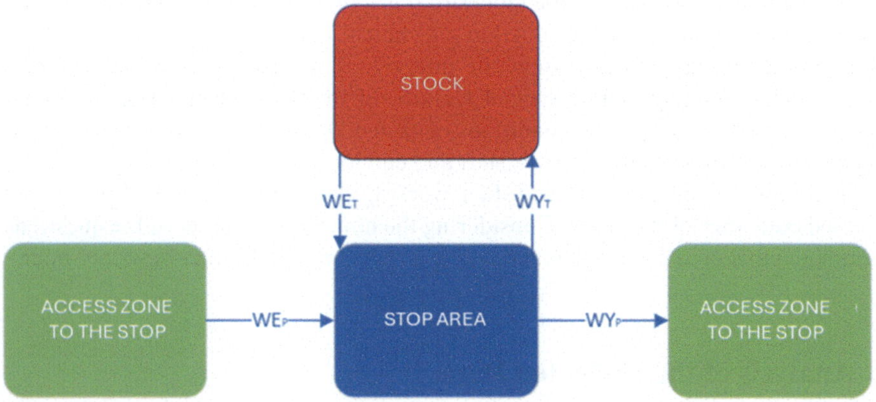

Fig. 8. Scheme of passenger exchange within the stop. WE_P—number of pedestrians who entered the stop, WE_T—number of passengers who entered the stop area by public transport, WY_P—number of pedestrians who left the stop, WY_T—number of passengers who left the stop area by public transport.

transport. It is important to note that not all individuals entering the stop area necessarily use public transport to leave. At transit stops, some individuals may enter and leave the stop without utilizing its transport function. The passenger exchange scheme at the stop is illustrated in Fig. 8.

3.3 Assessment of Cordon Tightness

The term "travelers' cordons" refers to the flows of travelers entering and exiting the stop area. The "tightness of cordons" indicates the consistency of measurements of passenger flows, specifically the number of people entering and leaving the stop area.

Cordons are considered tight if the number of travelers who entered the stop area (WEP + WET) equals the number of travelers who left the stop area (WYP + WYT) within a given time interval.

In the first time interval considered, the number of travelers who entered the stop area should be added to the number of travelers present in the stop area at the start of the study. Similarly, in the last time interval considered, the number of travelers who left the stop area should be added to the number of travelers remaining in the stop area at the end of the survey. If the measurement results are inconsistent, the relative error of the cordon tightness is obtained, which is determined by the formula (1):

$$B = \left|(WE_P + WE_T) - WY_P + WY_T)\right| \quad (1)$$

where:

B—relative error of cordon tightness
WE_P—the number of pedestrians who entered the stop area
WE_T—number of passengers who entered the stop area by public transport
WY_P—number of pedestrians who left the stop area
WY_T—the number of passengers who left the stop area by public transport

Any differences between the (and () totals may result from measurement errors in passenger counting. If possible, we use the method described in point 3.1 to verify the correctness of the measurements. If we are unable to obtain consistent results for the number of people entering and leaving the stop area, in In order to enable further analyses, the obtained cordon tightness errors should be compensated by subtracting 70% of the cordon's relative tightness error from the number of people and adding 30% of the cordon's relative tightness error to the number of people. This action is presented in the following formulas (2) and (3):

$$WE_{TK} = WE_T - 0,7 * B \quad (2)$$

$$WY_{TK} = WY_T + 0,3 * B \quad (3)$$

where:

WE_{TK}—number of passengers who entered the stop area by public transport after eliminating the cordon tightness error
WY_{TK}—the number of passengers who left the stop area by public transport after eliminating the cordon tightness error
WE_T—number of passengers who entered the stop area by public transport
WY_T—the number of passengers who left the stop area by public transport
B—relative error of cordon tightness

The values of WE_{TK} and WY_{TK} are used for further analyses.

3.4 Transfer Rate

An important functionality of transfer hubs is the ability for travelers to change their direction or means of transport. Measurement results at junctions can be utilized to determine the transfer rate, which indicates how many people use the junction's transfer function. To objectively assess the utilization of the junction's transfer function, the transfer rate is employed.

The transfer rate (WPR) is a standardized measure that expresses the percentage of people transferring relative to the total number of people using the stop. This is determined by the formula (8):

$$W_{PR} = \frac{PR}{U} * 100\% \tag{4}$$

where:

W_{PR}—transfer rate
PR—number of passengers changing transport modes at the stop
U—number of people using the stop

The number of people transferring (PR) is defined as the number of individuals who entered the stop area using collective public transport and subsequently left the stop area using collective public transport as well. This number is determined by formulas (4) and (5):

$$PR = WE_{TK} - (WY_P - P) \tag{5}$$

where:

PR—number of passengers changing transport modes at the stop
WE_{TK}—number of passengers who entered the stop area by public transport after eliminating the cordon tightness error
WY_P—number of pedestrians who left the stop area
P—number of people passing through the stop area (transit—subset of the component WY_P)

or

$$PR = WY_{TK} - (WE_P - P) \tag{6}$$

where:

PR—number of passengers changing transport modes at the stop
WY_{TK}—the number of passengers who left the stop area by public transport after eliminating the cordon tightness error
WE_P—number of pedestrians who entered the stop area
P—number of people passing through the stop area (transit—subset of the component WE_P)

Formulas 4 and 5 are alternatives to each other and, assuming the tightness of the cordons, lead to the same results. The key to determining the number of people transferring at a given junction and, therefore, the transfer rate is the identification at the stage of field research of the number of people passing (P) through the stop area (transit—a subset of components WE_P and WY_P).

The number of people using the stop (U) is determined by formulas (7) and (8):

$$U = WY_{TK} + WY_P - P \tag{7}$$

where:

U—the number of people using the stop
WY_{TK}—the number of passengers who left the stop area by public transport after eliminating the cordon tightness error
WY_P—number of pedestrians who left the stop area
P—number of people passing through the stop area (transit—subset of the component WY_P)

or

$$U = WE_{TK} + WE_P - P \tag{8}$$

where:

U—the number of people using the stop
WE_{TK}—number of passengers who entered the stop area by public transport after eliminating the cordon tightness error
WE_P—number of pedestrians who entered the stop area
P—number of people passing through the stop area (transit—subset of the component WE_P)

Formulas 6 and 7 are alternatives to each other and, assuming the tightness of the cordons, lead to the same results.

4 Study Sites

As part of the "Transfer without Barriers" project, pilot studies were conducted at two transfer hubs with different spatial and functional characteristics. The Gdańsk Świętokrzyska junction is a local hub primarily serving urban transport (tram and bus), while Gdańsk Śródmieście is a metropolitan hub dominated by rail transport (SKM).

The Gdańsk Świętokrzyska integration node is located on the periphery of Gdańsk's center but is one of the most significant transfer points for the southern districts of the city, which are predominantly residential. Access to the Gdańsk Świętokrzyska junction is primarily by bus, from which travelers can take a tram to the Main Railway Station in Gdańsk in 20 min or to the center of Wrzeszcz in 30 min. The development of satellite estates has attracted new investors, leading to a continuous increase in the number of people using the node.

The Gdańsk Śródmieście junction is situated in the central part of the city, at the intersection of Okopowa and Armii Krajowej streets. This node is of metropolitan significance, with rail transport (SKM) as the dominant mode, integrating rail, tram, and bus transport within its area. Major traffic generators near the junction include the Forum shopping mall, the Marshal's Office, and the Main Town with numerous tourist attractions.

At both nodes, the research methods described in Sect. 2 were employed, and the analyses outlined in Sect. 3 were conducted based on the obtained results. Due to space limitations, the article does not include a detailed description of the research methodology; instead, the most important results are discussed succinctly in the next section.

5 Results

The research findings on passenger flows reveal the presence of morning and afternoon peaks in the number of passengers utilizing the hubs. The onset and conclusion of these peaks exhibit some variations depending on the node's location. This phenomenon was observed through the adoption of a 5-min testing interval in the methodology. For the purpose of presenting the results, the data was aggregated into hourly intervals. At the Gdańsk Świętokrzyska junction, situated outside the city center, the morning peak commences slightly earlier compared to the Gdańsk Śródmieście interchange.

The age distribution of passengers and responses to survey inquiries indicate that during peak hours, the primary motivation for travel is commuting between home and work or school (refer to Fig. 9).

Moreover, pilot studies enabled the determination of the actual durations of transport peaks, which diverged from the initially assumed periods. This is distinctly illustrated in the example of traffic volume and parking lot occupancy at the Gdańsk Świętokrzyska junction (see Fig. 10).

The calculated pedestrian numbers reveal particularly congested routes or periods of heightened pedestrian activity. However, actual movements may not always align with

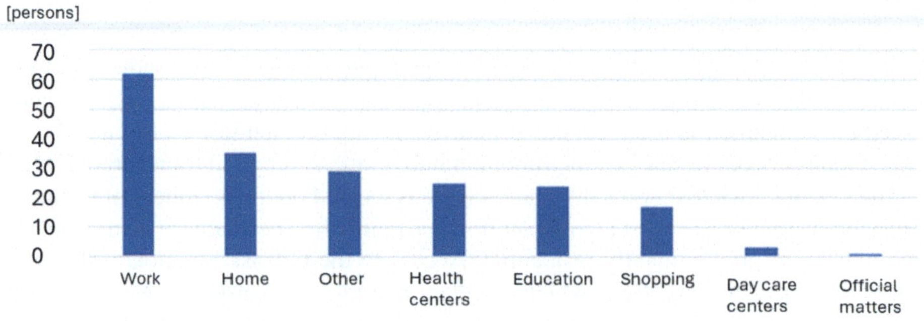

Fig. 9. Travel motivations of surveyed passengers.

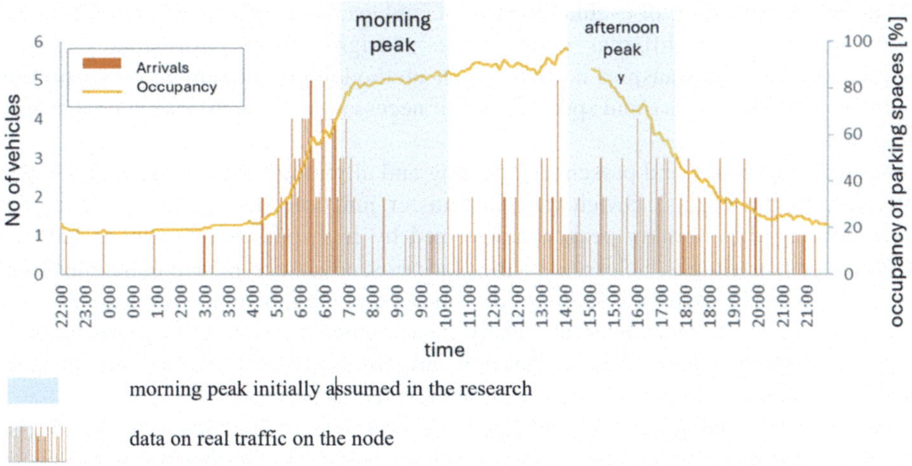

Fig. 10. Visualization of the measurement results of the number of cars entering the parking lot and the occupancy of spaces at the Gdańsk Świętokrzyska interchange.

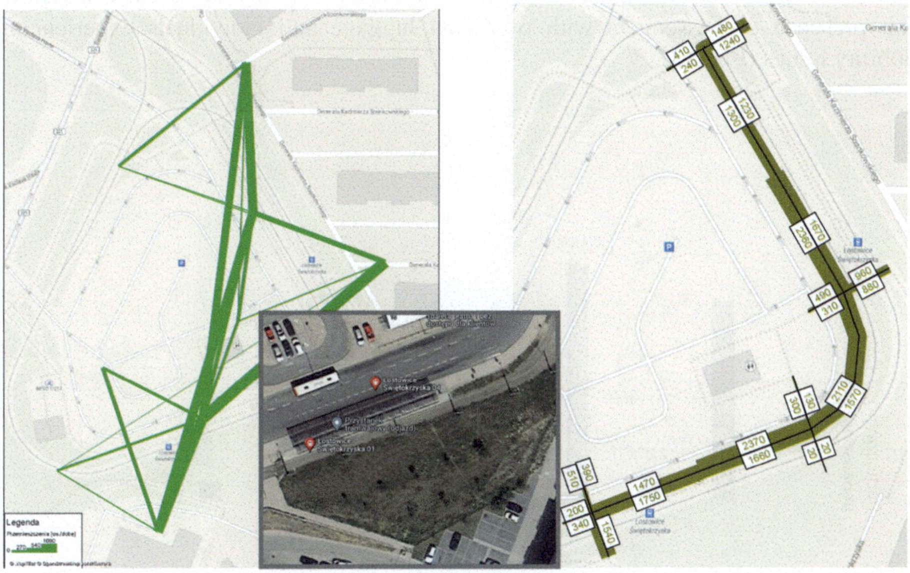

Fig. 11. Visualization of data from pedestrian traffic measurements at the Gdańsk Świętokrzyska interchange (cartogram) and satellite view of a trodden path.

existing infrastructure solutions. In cases where there is no designated pedestrian infrastructure for observed pedestrian traffic, spontaneous solutions emerge in real space, such as well-trodden paths (Fig. 11).

The research findings were directly utilized in the development of a method for assessing node accessibility. Preliminary results facilitated the creation of a node classification, identification of evaluation criteria, and establishment of minimum accessibility standards for different node classes. Insights from pilot studies guided modifications to the transport node research methodology, refinement of supporting tools (e.g., tables, sheets), and specification of necessary actions for subsequent stages, steps, and activities.

Vehicle occupancy and passenger boarding and alighting were scrutinized at 5-min intervals. At the Gdańsk Świętokrzyska transfer junction, the results reveal a brief period during the morning peak characterized by high passenger volumes both on arriving and departing vehicles and on the platforms, adversely impacting traveler comfort and safety. The dynamic nature of the situation at the junction was observable due to the short measurement intervals. Daily measurements enable the identification of problem intensity, frequency, and duration, providing vital information for management. To enhance readability, data can be aggregated as necessary.

Observations and interviews indicated the presence of individuals from diverse social and demographic groups among the passengers, including Persons with Special Needs (PwSN). The proportion of PwSN varies across different stops and modes of transport served (Fig. 12).

During the pilot studies, the predominant group among Persons with Special Needs (PwSN) consisted of individuals with strollers. Observers also documented a significant presence of passengers with oversized luggage and individuals experiencing mobility issues (Fig. 13).

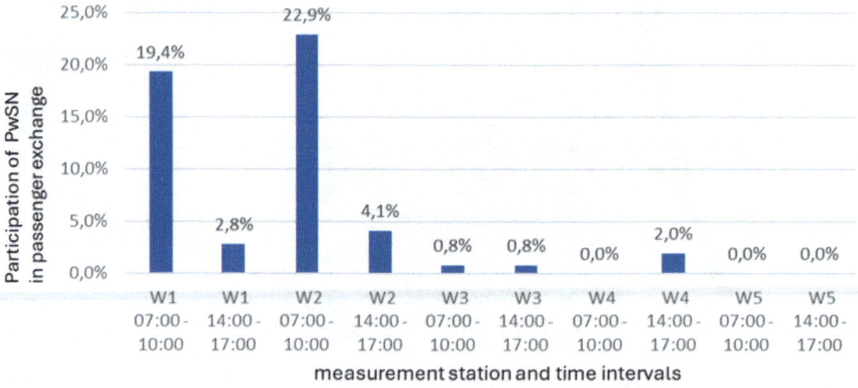

Fig. 12. Visualization of the time-varying share of PwSN at one of the stops at the Gdańsk Świętokrzyska interchange.

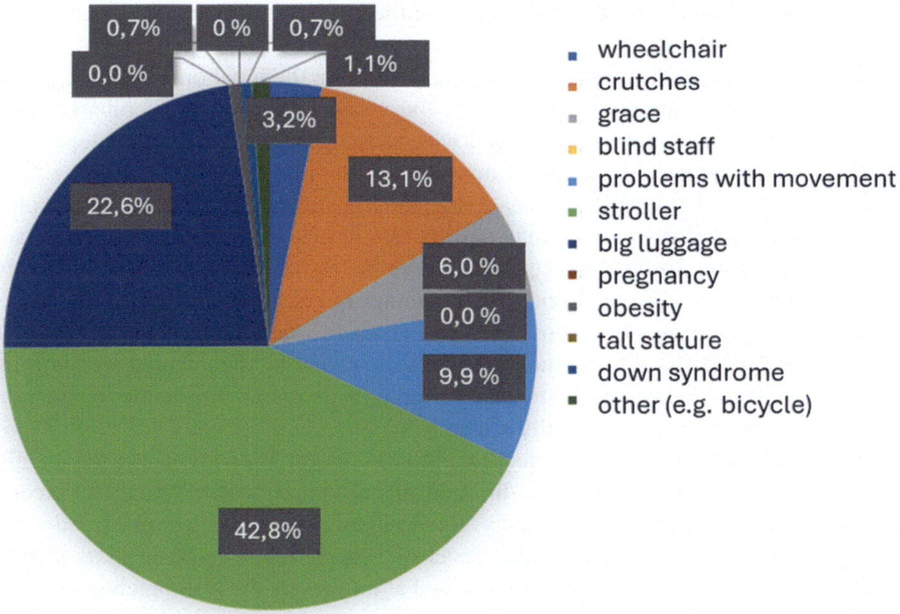

Fig. 13. Visualization of the results of research on the share of PwSN according to the adopted groups among people boarding and alighting in the morning and afternoon peak at the Gdańsk Świętokrzyska interchange.

6 Discussion

Interchange transport nodes serve as pivotal points in the urban transport network, facilitating the interchange of passengers between different modes of transport and/or altering their travel direction. The significance of these nodes within the city's transportation system is contingent upon their geographical positioning within the urban structure and the specific subsystems they incorporate.

The passenger traffic load exhibits temporal variations, with daily schedules typically encompassing morning and afternoon peaks. The temporal intervals of these peak occurrences serve as a distinctive attribute for a given node and are influenced by various factors, including its geographical location within the urban structure.

The passenger demographic encompasses individuals with special needs, referred to as the OzSP group. The proportion of this group may fluctuate across nodes, transport modes, and even individual stops within a single node or mode of transport. Predominantly, the PwSN group, particularly those with mobility impairments, exhibit a higher propensity to utilize tram or bus transportation over rail services.

In the nodes analyzed, the PwSN groups were predominantly composed of individuals with strollers, alongside a significant proportion of passengers with oversized luggage and mobility impairments. However, the absence of wheelchair users, particularly in the context of Gdańsk, may be attributed to the provision of a dedicated transport system for individuals with disabilities within the city. This scenario may vary across different urban contexts.

Transport studies yield a plethora of data essential for the functional characterization of nodes. Research outcomes pertaining to the volume of passengers embarking and disembarking transport services, coupled with passenger flow data within stops, serve as critical input parameters for the computation of the transfer rate (denoted as 2.8 in the original text). However, the evaluation of the transfer rate poses a significant challenge and may not always be feasible, particularly in the context of spatially extensive or multi-tiered interchanges. In scenarios where pedestrian flows are not comprehensively captured by recording devices (e.g., tunnel intersections, blind spots between adjacent cameras), it may be impossible to ascertain the variable from the transfer rate formula that determines the volume of individuals traversing the interchange/platform/stop (P). The computation of the transfer rate may be rendered unfeasible in situations where the cordon is not hermetically sealed.

Passenger traffic studies conducted over brief temporal intervals enable the identification of daily occurrences and the duration of individual conflict situations precipitated by high pedestrian traffic density. Such situations adversely impact pedestrian traffic conditions, as indicated by the Level of Service (LOS). The insights gleaned from these studies can inform strategies aimed at mitigating the phenomenon of excessive spatio-temporal concentration of passengers, such as timetable modifications. Consequently, the findings of transport research serve as a valuable resource for managers of transport interchange nodes. The test results from the Gdańsk Świętokrzyska node and the Gdańsk Śródmieście have been submitted to the Roads and Greenery Authority in Gdańsk.

The research findings were utilized during the development of a general methodology for conducting an audit of the availability of existing Interchange Transport Nodes (ITNs). Accordingly, the studies were conducted for nodes with diverse characteristics, encompassing spatial, functional, and types of ITN. The Gdańsk Świętokrzyska interchange is a local node characterized by dominant urban transport modes (tram, bus), while Gdańsk Śródmieście is a metropolitan interchange, with rail transport (SKM) being the dominant mode. For comprehensive representation of all node types, the following locations are recommended for further research due to their geographical scope:

- Lech Wałęsa Airport—an international interchange, predominantly serviced by air transport.
- Tczew Główny—a national interchange, primarily serviced by long-distance bus transport
- Gdańsk Wrzeszcz—a regional interchange, predominantly serviced by rail transport.

References

1. Act of July 19, 2019 on ensuring accessibility for people with special needs, Dz.U. 2019 poz. 1696
2. Bhalla, A., Lapeyre, F.: Social exclusion: towards an analytical and operational framework. Dev. Change. **28**(3), 413–433 (1997)
3. Okraszewska, R., Jamroz, K., Wachnicka, J., et al.: Managing the availability of Integrated Transfer Nodes, Transport in the era of diverse needs. In: Sitarz, M. (ed.) Transfers Without Barriers, pp. 105–126. Akadamie WSB, Dąbrowa Górnicza (2023) (in Polish)
4. Kamruzzaman, M., Yigitcanlar, T., Yang, J., Mohamed, M.A.: Measures of transport-related social exclusion: a critical review of the literature. Sustainability. **8**, art. 696 (2016)
5. Sitarz, M.: Introduction to: Transport in the era of diverse needs. In: Sitarz, M. (ed.) Transfers Without Barriers, pp. 105–126. Akademia WSB, Dąbrowa Górnicza (2023) (in Polish)
6. Jamroz, K., Okraszewska, R., Sitarz, M., Zieja, M., et al.: Availability of transport interchanges for people with special needs. Ann. Traffic Eng. Transp. Plann. **4**(XIII), 363–380 (2021) Horizon 2050 Better Transport & Better City, Poznań (in Polish)
7. Okraszewska, R., Jamroz, K., Wysocki, M., Szmagliński, J.: Draft accessibility standards for people with special needs at transport interchanges. Ann. Traffic Eng. Transp. Plann. **5**(XIV), 129–143 (2023) Traffic Planning and Global Challenges/New Means of Transport and New Forms of Mobility, Poznań (in Polish)

Impact of Changing Bus Propulsion Systems Based on EU-Funded Projects on the Total Cost of Ownership and Social Evaluation of Urban Transport Travel—Analysis Results

Edyta Boratyńska Karpiej[1] and Paweł Engel[1] (✉)

[1] Department of Transport Analysis, Centre for EU Transport Projects, Warsaw, Poland
eboratynska-karpiej@cupt.gov.pl; pengel@cupt.gov.pl

Abstract. This article presents a summary of an analysis, conducted by a Research Team (The Team consisted of CEUTP experts Edyta Boratyńska-Karpiej, Dariusz Giziński, Anna Kiwiel, Marta Lorczyk. Expert supervision was exercised by Paweł Engel) from the Centre for EU Transport Projects, on the total cost of ownership and efficiency (TCO) of urban buses. The study was based on projects funded by European Union (EU) under the Operational Programme Infrastructure and Environment 2014–2020. It comes as no surprise that the value of the ownership and operating costs of urban buses depends on the type of propulsion system. Electric-powered vehicles, for instance, require batteries to be replaced during the life of the bus. Operators considering the purchase low-emission vehicles should factor in the cost of additional infrastructure for electric buses which often includes adapting depots or bus stops to install charging facilities. The TCO provides an estimate of the total vehicle costs associated with purchase, use and decommissioning over the entire life cycle.

In addition to the TCO, public survey was conducted to gather citizens' views on the changes in urban bus propulsion systems funded by the EU.

Keywords: Transport · Urban transport · Propulsion · Cost · Evaluation · Travel comfort · Zero-emission · Electromobility

1 Introduction

Public transport is essential for societal development, urban mobility, and economic efficiency. In view of global climate change, increasing importance is attributed to the implementation of innovative solutions that reduce emissions and fossil fuel consumption. In Europe, particularly in Poland, transport is one of the key sectors of the economy that requires a transformation towards zero-emission [1].

The report, prepared by the Centre for EU Transport Projects (CEUTP), analyses the impact of changing the propulsion of urban buses as part of urban public transport, using data obtained from EU-funded projects under the Operational Programme

G. Sierpiński et al. (Eds.): TSTP 2024, LNNS 1257, pp. 199–206, 2024.
https://doi.org/10.1007/978-3-031-81777-9_13

Infrastructure and Environment. It also presents methods for assessing the total cost of ownership (TCO) of different bus types and their environmental and social impacts.

2 Analysis Objectives

The main objective of the analysis was to estimate the total costs of owning and operating urban buses with different propulsion types—conventional, gas, hybrid, electric and hydrogen. The analysis considered not only the purchase costs, but also those associated with fuel or energy consumption, repairs, battery replacement and the construction of charging and refueling infrastructure.

This information can help public transport operators make informed decisions about investments in low-carbon transport technologies that will have long-term benefits for urban air quality and noise reduction. The conclusions of the analysis also aim to support the European Union's strategic action towards achieving climate neutrality targets by 2050, in line with the "Fit for 55" initiative [2].

2.1 Actuality

Climate change is one of the greatest challenges facing the world today. Global warming is contributing to an increase in the frequency of extreme weather events, which in turn has a huge impact on the economies of European countries. According to desk research analysis, between 1980 and 2020, the damage caused by climate phenomena in Europe amounted to more than EUR 487 billion and the number of fatalities exceeded 138,000 people [1].

In the context of these challenges, electromobility, i.e. the development of transport using electric vehicles, is becoming one of the priorities for environmental and economic action in the European Union. As a member of the EU, Poland is implementing a number of programmes and projects aimed at developing infrastructure and popularising low-emission vehicles [1].

2.2 Formal and Legal Conditions

The development of electromobility and the use of alternative fuels in Poland is regulated by a number of legal acts, including the 2018 Act on Electromobility and Alternative Fuels [3]. This act defines the rules for the development and operation of alternative fuel infrastructure and the obligations of public entities in its development.

As part of EU climate policy, Member States, including Poland, have committed to reducing greenhouse gas emissions and increasing the share of renewable energy in transport. Projects co-financed from EU funds under the Infrastructure and Environment Operational Programme, among others, contribute to the achievement of these goals by modernising the bus fleet in Polish cities and building charging infrastructure for electric vehicles.

2.3 Alternatives to Diesel Engines

European climate policy places numerous obligations on member states to reduce CO_2 emissions [2]. As a result, it is necessary to look for alternatives to traditional petroleum-based fuels. The report of the impact of changing propulsion systems in urban transport buses based on EU-funded projects [1] discusses several alternative energy sources for transport, including electricity, hydrogen, natural gas (CNG, LNG) and hydrogenated vegetable oil (HVO).

Electricity
Electric buses, which draw their energy from external power sources, are the most popular alternative to internal combustion engines. Despite higher purchase costs, electric buses have lower operating costs, particularly in terms of energy consumption and repair.

Hydrogen
Hydrogen technology is a promising alternative to electric vehicles, especially in the context of longer routes. Hydrogen can be used as a fuel to power buses, helping to reduce atmospheric emissions.

Natural Gas (CNG and LNG)
Buses powered by compressed (CNG) or liquefied natural gas (LNG) are another alternative to traditional fossil fuels. CNG and LNG emit less harmful substances compared to diesel, but still require further expansion of refuelling infrastructure.

3 Total Cost of Ownership (TCO) Analysis

A study on the cost of ownership of buses with different types of propulsion, including EU-funded vehicles, was carried out in business units from 30 cities of different sizes and nature of operation. The aim of the study was to compare the total cost of ownership (TCO) of diesel, CNG, hybrid, electric and hydrogen buses. The results included operating and investment costs, which were collected based on data from public transport units from 2023. Vehicles were evaluated based on cost per vehicle kilometre, which allowed for comparison regardless of differences in fleet maintenance patterns.

3.1 Distribution of Propulsion Systems in the Survey

A total of 72 respondents took part in the survey, sharing data on the use of the following propulsion types:

- Diesel: 30 respondents;
- CNG: 4 respondents;
- Hybrid: 13 respondents;
- Electric with charging at the depot: 6 respondents;
- Electric with rapid pantograph charging: 16 respondents;
- Hydrogen: 3 respondents.

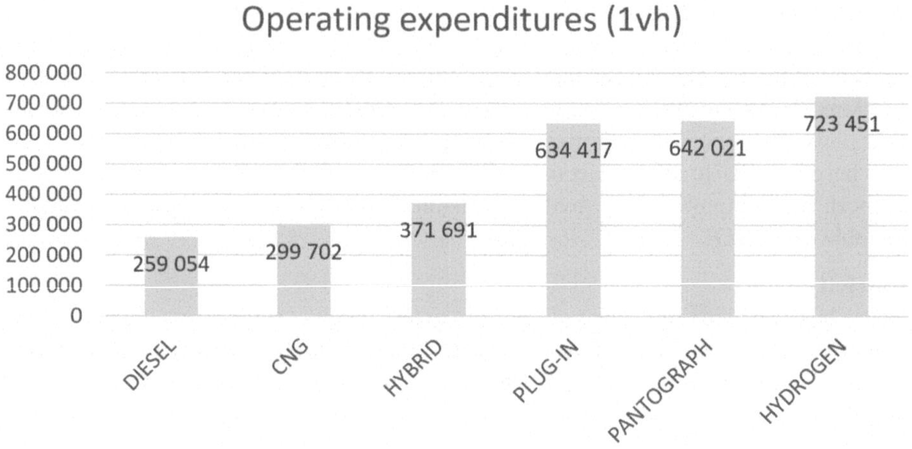

Fig. 1. Operating expenditures. Source: Own study based on [1].

Over their lifetime, hydrogen buses proved to be the most expensive to operate, due to the high cost of fuel and the infrastructure required to operate them. These costs were further exacerbated by the small number of hydrogen buses in service, limiting the ability to spread the expenditure over a larger fleet (Fig. 1).

With an average annual mileage of 60,000 km, electric buses charged only at the depot were found to be the cheapest, followed by hybrid buses. For Diesel, CNG and electric buses with fast pantograph charging, costs were similar, but the highest costs were recorded for CNG and Diesel. It is worth noting that at higher annual mileage (80,000 km), electric buses become significantly more cost-effective compared to diesel vehicles.

Service and Repair Costs

The highest repair and servicing costs were recorded for Diesel buses, due to the more complex mechanism of these vehicles. Electric and hydrogen buses have lower service costs due to the simpler mechanical design of electric drives.

Fuel and Energy Costs

The energy cost of the vehicles analysed varied. The highest fuel costs were recorded for hydrogen and CNG buses, while the energy cost for electric buses was stable and predictable. The diesel price was assumed to be 4.46 PLN net per litre. However, fuel and energy costs may fluctuate depending on the energy market situation and developments in production technology, especially for hydrogen.

Electric Bus Operation

Electric buses charged at the depot have proven to be the cheapest option in terms of cost per vehicle kilometre, especially for low and medium mileage. However, for vehicles using pantograph fast chargers, infrastructure costs significantly increase the total cost of ownership, especially for lower annual mileage.

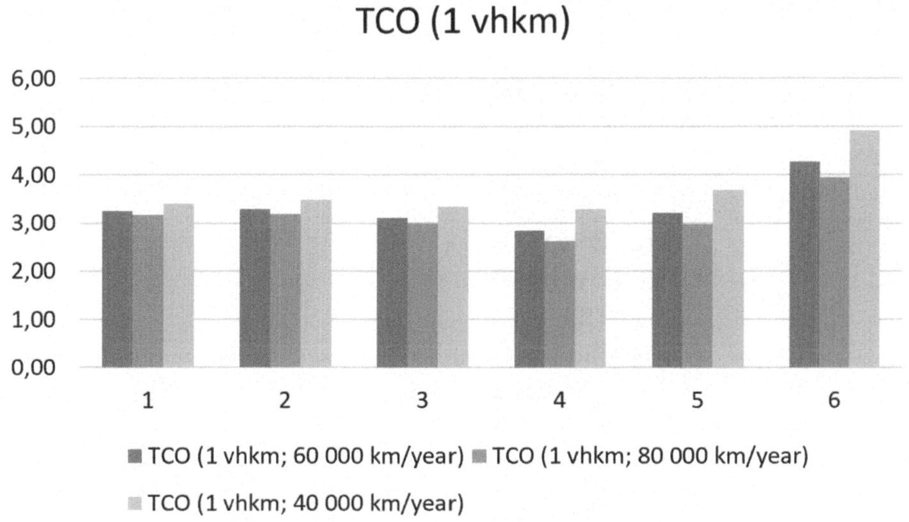

Fig. 2. TCO. Source: Own study based on [1].

Due to the early stage of deployment and low annual mileage (less than 30,000 km on average), hydrogen buses had the highest cost of ownership per vehicle kilometre. However, it is projected that as hydrogen production technology develops and production scales up, these costs could decrease significantly in the future (Fig. 2).

An analysis of the cost of ownership of buses with different propulsion systems showed that choosing the right propulsion system depends on a number of factors, including infrastructure, annual mileage and fuel or energy availability. Electric buses have proven to be the most cost-effective, especially for medium and high mileage, while hydrogen buses, although the most expensive at present, have potential for the future due to technology developments and falling hydrogen prices. The running costs of internal combustion drives, especially Diesel, remain high due to maintenance and repair costs, which may influence the further development of electromobility in the public transport sector.

Public Acceptance of Change

A survey on public acceptance of EU-funded changes to urban rolling stock shows that Polish citizens are increasingly aware of the need for modern transport solutions. The aim of the survey was to gather information on urban residents' views on changes to urban bus propulsion systems and their knowledge of alternatives to traditional diesel engines. The survey involved 670 people, representing different age groups, living in both large cities and smaller towns.

The survey showed that the most common means of transport chosen by respondents was the traditional diesel car (38%). Nevertheless, as many as 30% use public transport and 26% travel by foot or bicycle. The frequency of bus use also varied, with a sizable proportion of respondents using this mode of transport on a daily basis. Diesel, electric and hybrid buses were the most popular vehicles among those who used urban buses (Fig. 3).

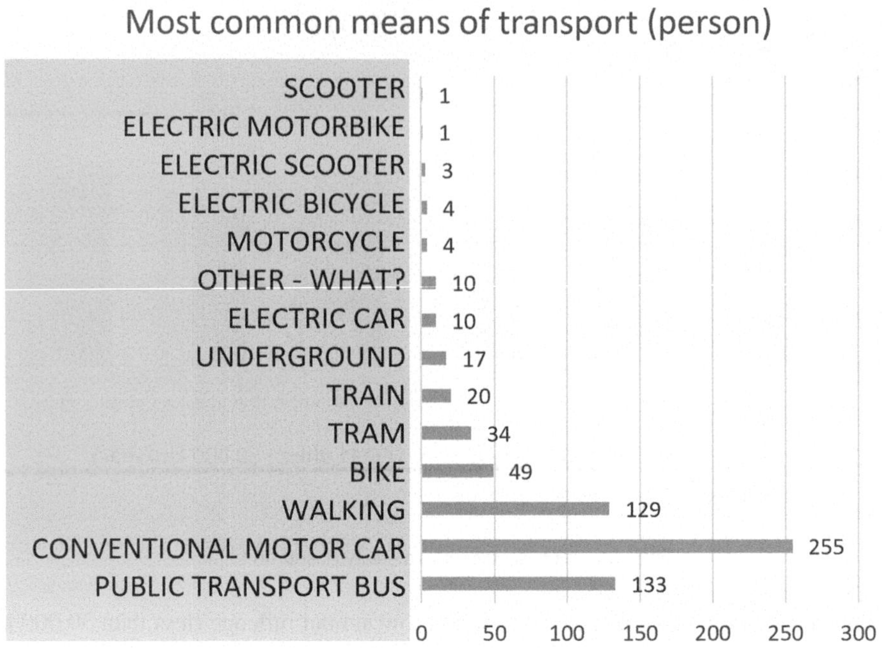

Fig. 3. Most frequently used means of transport by respondents. Source: Own study based on [1].

In terms of switching from traditional cars to public transport, the majority of car owners showed no willingness to change their transport habits. Only 24% expressed a willingness to consider such a change and as many as 46% categorically rejected it. A significant proportion of people (30%) were undecided, suggesting that education and public campaigns could be effective in changing their attitudes.

The survey also highlighted that for many respondents, electromobility represents the future of motoring. As many as 55% of respondents expressed the belief that electric drive technologies will play a key role in the development of transport. Additionally, 66% of respondents noted that zero-emission transport can effectively contribute to reducing smog in cities, an important step in the fight for better air quality.

In the context of the evaluation of changes to date, 76% of respondents believe that EU funds should continue to support the replacement of urban fleets with zero-emission vehicles. A significant proportion of people (72%) stressed that state support is key to accelerating this process. Interestingly, 41% of those who experienced an improvement in travel comfort in recent years attributed this change to the use of modern powertrains in buses (Fig. 4).

In summary, the results of the survey show that despite the widespread use of traditional cars, the public recognises the need for changes towards electromobility and the modernisation of public transport. However, further education and financial support, both at national and EU level, are needed to successfully implement these changes.

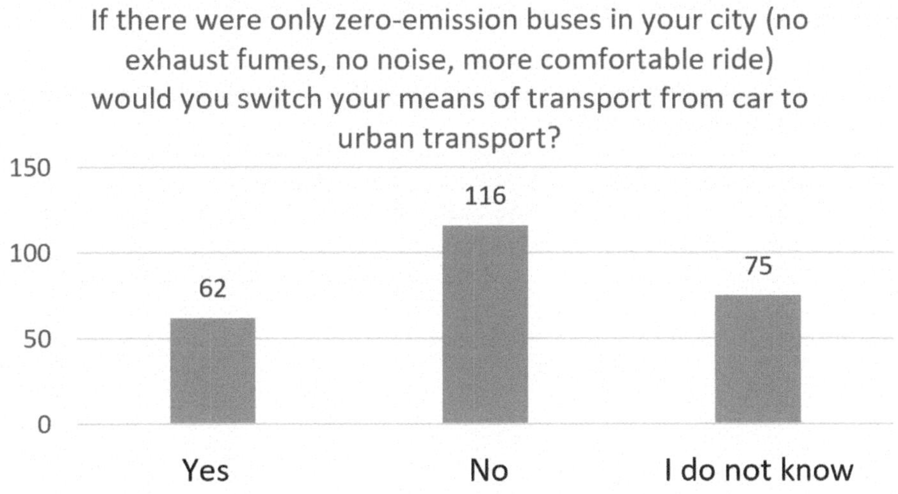

Fig. 4. Potential change from car to public transport. Source: Own study based on [1].

4 Conclusions

EU funding programmes support the development of modern infrastructure and encourage cities to invest in green transport. In terms of the financial aspect of vehicle operation, the total cost of ownership, considering current cost levels, during the life of 15 years is equalised for all types of bus propulsion, excluding hydrogen. This calculation does not include differences in external costs, which will in principle be lower for zero-emission vehicles.

The CEUTP report indicates that continuing efforts to change the propulsion systems of urban buses is possible and is not a more expensive alternative to traditional vehicles in the long term. Modern propulsion technologies and financial support from the EU are the foundation for the transformation of public transport in Poland.

In order to reduce the further external costs of zero-emission transport, it is necessary to conduct activities related to the recycling and reuse of batteries, as well as to minimise the use of components of propulsion systems that are difficult to obtain, including so-called rare earth elements.

The entire report is available on the CEUTP website: www.cupt.gov.pl. With the study we wanted to start a discussion on the future of zero-emission in urban transport. We encourage comments on the CEUTP's conclusions: badanieautobusowe@cupt.gov.pl. They will allow us to continue the analysis and revise the conclusions.

References

1. CEUTP: Analysis of the impact of changing propulsion systems in urban transport buses based on EU-funded projects (2024)
2. European Commission: The European Green Deal. https://commission.europa.eu/strategy-and-policy/priorities-2019-2024/european-green-deal_en
3. Act on Electromobility and Alternative Fuels. https://sip.lex.pl/akty-prawne/dzu-dziennik-ustaw/elektromobilnosc-i-paliwa-alternatywne-18683445

Author Index

G. Sierpiński et al. (Eds.): TSTP 2024, LNNS 1257, p. 207, 2024.
https://doi.org/10.1007/978-3-031-81777-9

SPRINGER NATURE

GPSR Compliance

The European Union's (EU) General Product Safety Regulation (GPSR) is a set of rules that requires consumer products to be safe and our obligations to ensure this.

If you have any concerns about our products, you can contact us on ProductSafety@springernature.com

In case Publisher is established outside the EU, the EU authorized representative is:

Springer Nature Customer Service Center GmbH
Europaplatz 3
69115 Heidelberg, Germany

The manufacturer's authorised representative in the EU is Springer Nature Customer Service Centre GmbH, Europaplatz 3, 69115 Heidelberg, Germany. If you have any concerns regarding our products, please contact ProductSafety@springernature.com

Printed and bound by CPI Group (UK) Ltd, Croydon, CR0 4YY
01/10/2026
02225911-0001